DATE DUE		
MAY 0 7 '98		
AUG 6 '98		
MAR 1 6 2000		
MAR 2 7 2000		
AUG 0 1 2000		
DEC 1 8 2000		
NOV 3 0 2001		
DEC 1 3 2001		

About Island Press

Island Press, a nonprofit organization, publishes, markets, and distributes the most advanced thinking on the conservation of our natural resources—books about soil, land, water, forests, wildlife, and hazardous and toxic wastes. These books are practical tools used by public officials, business and industry leaders, natural resource managers, and concerned citizens working to solve both local and global resource problems.

Founded in 1978, Island Press reorganized in 1984 to meet the increasing demand for substantive books on all resource-related issues. Island Press publishes and distributes under its own imprint and offers these services to other nonprofit organizations.

Support for Island Press is provided by The Geraldine R. Dodge Foundation, The Energy Foundation, The Charles Engelhard Foundation, The Ford Foundation, Glen Eagles Foundation, The George Gund Foundation, William and Flora Hewlett Foundation, The James Irvine Foundation, The John D. and Catherine T. MacArthur Foundation, The Andrew W. Mellon Foundation, The Joyce Mertz-Gilmore Foundation, The New-Land Foundation, The Pew Charitable Trusts, The Rockefeller Brothers Fund, The Tides Foundation, and individual donors.

About the Natural Resources Defense Council

NRDC is a nonprofit membership organization with more than 170,000 members and contributors nationwide. Since 1970, NRDC scientists and lawyers have been working to safeguard the Earth: its people, its plants and animals, and the natural systems on which all life depends. NRDC works to restore the integrity of the elements that sustain life—air, land, and water—and to defend endangered natural places. We seek to establish sustainability and good stewardship of the Earth as central ethical imperatives of human society.

NRDC pursues its goals primarily by shaping public policy and private economic activity. We work to craft innovative and pragmatic solutions to critical environmental problems. To put these solutions into practice and ensure their effectiveness, we advocate, negotiate, litigate, lobby, and educate. NRDC's major programs are in the areas of air, water, and land resources; human health; energy efficiency and renewable energy; the global environment; nuclear weapons proliferation and nuclear waste disposal; and the urban environment.

NRDC is the only environmental organization with a full-time staff committed to achieving the national goal of the Clean Water Act: ending the discharge of wastes into the nation's lakes, rivers, and coastal waters. We are grateful to the following donors for their generous support of our water quality work, of which this book is a part: Mary Owen Borden Memorial Foundation, Carolyn Foundation, Geraldine R. Dodge Foundation, Chichester DuPont Foundation, The Joyce Foundation, The McKnight Foundation, The Prospect Hill Foundation, Reese Family Fund of the Boston Foundation, Steven C. Rockefeller, Town Creek Foundation, Victoria Foundation, and Virginia Environmental Endowment. In addition, we thank the 170,000 members of NRDC, whose support makes our water quality work possible.

The Clean Water Act
20 Years Later

The Clean Water Act
20 Years Later

**Robert W. Adler, Jessica C. Landman,
and Diane M. Cameron**

 Natural Resources Defense Council

ISLAND PRESS

Washington, D.C. ■ Covelo, California

To Woody, Merrill, Rebecca, and their generation.

Library of Congress Cataloging-in-Publication Data
Adler, Robert W., 1955–
    The Clean Water Act 20 years later / Robert W. Adler, Jessica C.
Landman, and Diane M. Cameron.
        p.   cm.
    Includes bibliographical references and index.
    ISBN 1-55963-265-8 (cloth : alk. paper) : ISBN
1-55963-266-6 (pbk. : alk. paper)
    1. Water quality—United States.   2. Water quality management—
United States—History.   3. United States. Federal Water Pollution
Control Act.   I. Landman, Jessica C.   II. Cameron, Diane M.
III. Title.   IV. Title: Clean Water Act twenty years later.
TD223.A4663   1993
363.73′946′0973   93-22632
                                                                                          CIP

Acknowledgments

The authors gratefully acknowledge the support and assistance of the donors to NRDC's clean water work (who are listed at the beginning of this book) and of the members, Board, management and staff of NRDC. We are particularly indebted to our Assistant, Carol James, who worked as hard as any of us to make this book a reality. Portions of this book were written or produced with the assistance of, or drawn on earlier work by, a large number of our current and past colleagues at NRDC and in the Clean Water Network. Marni Finkelstein helped draft parts of chapter 3. With apologies to those we may have excluded, we also thank David Bailey, Chris Calwell, Sarah Chasis, Richard Cohn-Lee, Ken Cook, David Dickson, Stephanie Grogan, Carolyn Hartmann, Jim Hecker, Andy Hug, Doug Inkley, Kathrin Day Lassila, Christophe Lawrence, Ashley McLain, Robbin Marks, Dawn Martin, Beth Millemann, Kailen Mooney, Steve Moyer, Erik Olson, Paul Orum, Robyn Roberts, Nina Sankovitch, Abby Schaefer, Denise Schlener, Greg Shaner, Debbie Sheiman, Sarah Silver, Jim Simon, Lisa Speer, Nancy Vorsanger, Justin Ward, and Clark Williams. We also appreciate the thoughtful comments provided by Frances Dubrowski, Dr. Jeffrey Foran, Dr. Mike Hirshfield, Dr. James Karr, Ann Powers, Mark Van Putten, and Linda Winter. Finally, we are indebted to the staff at Island Press, including Chuck Savitt, Joe Ingram, Beth Beisel, and Christine McGowan, and to our copy editor, Barbara Fuller, and our typesetter, Elyse Chapman.

Contents

Contents

Preface

Why a Book on the Clean Water Act?

In March 1990, U.S. Senators Max Baucus of Montana and John Chafee of Rhode Island introduced S. 1081, an omnibus bill to amend and reauthorize the federal Clean Water Act. The Clean Water Act was originally passed in 1972 and has been amended several times since, most notably in 1977 and 1987. With the introduction of S. 1081, serious debate began on the efficacy of our national programs to eliminate water pollution and to restore the health of our aquatic ecosystems.

Our efforts on behalf of the Natural Resources Defense Council (NRDC) to support Clean Water Act reauthorization began with a serious evaluation of how well the Clean Water Act has achieved its primary goal of restoring and protecting the integrity of the nation's surface waters. To our surprise, little work had been done on this subject. As we approached the twenty-year anniversary of this landmark law, no comprehensive analysis was available to answer basic questions: How much cleaner are our rivers than they were two decades ago? Are our coastal beaches safer for swimming? Do our lakes support more fish, and are the fish safer to eat? What is happening to waterfowl and other animals that rely on aquatic habitat?

Stacks and stacks of government reports were available on virtually all aspects of Clean Water Act program administration. Government computers were laden with megabytes of technical data, much of it designed in principle to answer basic questions about the quality of our surface waters and the health of our aquatic ecosystems. Most experts agreed, however, that these data were riddled with gaps and inconsistencies, making it almost impossible to determine whether, on a national basis, we have made significant progress in the war against water pollution. The program documents answered questions such as how many permits have been written, but told little about how well those permits controlled pollution or complied with the law.

Indeed, by the twentieth anniversary of the law, no one had written a meaningful, comprehensive analysis of the successes and failures of the Clean Water Act on a national scale, much less in a form understandable to the public or even reasonably educated citizen activists and policymakers. Technical reports and articles have been written critiquing implementation of specific programs and provisions of the law. And excellent books have been written exploring the health of individual water bodies, such as the Chesapeake Bay and the Great Lakes. Nothing was available on a national scale to answer the fundamental questions most people and public officials want answered about clean water.

We set out to fill this gap. Rather than rely only on traditional federal and state Clean Water Act program documents, which the experts agreed were of only limited value in responding to our basic questions, we searched for real-world information on the safety of our waters for swimming, fishing, and drinking; on the health of our aquatic ecosystems and the availability of important aquatic habitat; and on the status of fish and other species that depend on our rivers, lakes, and wetlands.

This search gave us some cause for hope that the United States has succeeded in reducing some forms of serious water pollution, especially from traditional sources such as factories and sewage treatment plants—so-called point sources of pollution. These gains have reduced the severity of fish kills; improved water quality in some regions; lowered toxic contamination of fish, shellfish, and other species in many waters; and made many beaches safer for swimming. But the best available data show that we still have a long way to go to reduce the serious impacts even from these point sources of pollution, which have received the largest amount of attention. Beaches remain unsafe for swimming, fishing bans and advisories remain in effect all over the country, and severe pollution incidents continue to kill millions of fish.

More important, we found little attention to the overall health of aquatic ecosystems. Because of massive polluted runoff from farms, city streets, and other intensive land uses (known to bureaucrats as *nonpoint source pollution*), as well as large-scale destruction of wetlands, floodplains, stream channels, coastlines, and other important aquatic habitat, we are actually going backward in our efforts to restore the health of our aquatic ecosystems. As a result, we are losing many of the aquatic resources we rely on for food, drinking water, jobs, and, in many cases, a way of life. Widespread media attention to land-dwelling animals such as spotted owls and red squirrels is well justified, but far more aquatic species are jeopardized than their terrestrial cousins. Even where not directly in danger of extinction, many key species of recreational and commercial fish and shellfish, waterfowl, and other aquatic-dependent species have declined severely since the 1972 Clean Water Act was passed.

We didn't ignore the many available government reports on program implementation. Having identified the major areas of progress and failure in meeting the real-world goals of the Clean Water Act, we turned next to a careful analysis of how well the U.S. Environmental Protection Agency (EPA) and the states have implemented the various programs and provisions of the law designed to achieve these goals. Again, our review showed significant progress in some areas and in some programs. Overall, however, we found that Clean Water Act programs remain riddled with serious gaps, loopholes, and other problems. Some of these result from inadequate funding and personnel, others from poorly written and sometimes illegal regulations, and others simply from a lack of will to enforce the law properly.

This book closes with a national agenda for clean water, NRDC's legislative agenda for Clean Water Act reauthorization, drawn from our own analysis of problems with the law and from reauthorization recommendations adopted by collaborative organizations such as the Clean Water Network and Water Quality 2000. We hope these recommendations will contribute to the ongoing national discussion of what steps are needed to achieve the goals we set as a nation in 1972—to restore the chemical, physical, and biological integrity of the nation's waters.

Part I
A Clean Water Retrospective

In 1965, President Lyndon B. Johnson condemned the quality of the Potomac River as part of his pledge of "Clean Water by 1975," and in 1969, a conference in Washington, D.C., declared the river to be "a severe threat to anyone who comes in contact with it." The modern war against air pollution began in 1970 with passage of the Clean Air Act. But the battle against water pollution remained in the nineteenth century, with an obscure law called the Rivers and Harbors Act of 1899 as its principal, badly outdated weapon.

President Richard Nixon ushered in the new decade by reminding the nation that "the 1970s absolutely must be the years when America pays its debt to the past by reclaiming the purity of its air, its waters, and our living environment. It is literally now or never." The statement gave hope that the federal government would begin to take the war against water pollution seriously.

More visible political events, however, overshadowed President Nixon's commitment to the environment. In an effort to honor his 1968 pledge to end the war, Nixon escalated the bombing of North Vietnam and Cambodia. While the nation continued to debate the morality and the wisdom of the bombing and the war itself, an ever-present subtheme was the cost of the war to the U.S. economy and the federal budget and its impact on pressing domestic problems. Could the United States fight communism in Vietnam and still afford to fight pollution at home?

It was precisely this economic trade-off that led President Nixon to veto the Clean Water Act. Nixon's veto message reiterated his rhetorical pledge to address water pollution but rejected Congress's solutions on economic grounds:

> I am also concerned, however, that we attack pollution in a way that does not ignore other very real threats to the quality of life, such as spiraling prices and increasingly onerous taxes. Legislation which would continue our efforts to raise water quality, but which would do so through extreme and needless overspending, does not serve the public interest. There is a much better way to get this job done.[1]

It took Congress only one day to override Nixon's veto, by overwhelming, bipartisan margins in both houses.[2] In the U.S. Senate, the most eloquent response came from Senator Ed Muskie of Maine, who led the fight for a more serious national water pollution control effort:

> Can we afford clean water? Can we afford rivers and lakes and streams and oceans which continue to make possible life on this planet? Can we afford life itself? Those questions were never asked as we destroyed the waters of our Nation, and they deserve no answers as we finally move to restore and renew them. These questions answer themselves. And those who say that raising the amounts of money called for in this legislation may require higher taxes, or that spending this much money may contribute to inflation simply do not understand...this crisis.[3]

Nor did partisan politics stand in the way of the Clean Water Act. Senator Howard Baker of Tennessee, the ranking Republican member of the Senate Environment Committee, responded:

> I am deeply disappointed that President Nixon has chosen to veto the [act]. I hope that my colleagues will vote to override the President's veto....

> I believe that the [act] is far and away the most significant and promising piece of environmental legislation ever enacted by Congress.... Of course such an ambitious program will cost money—public money and private money. The bill vetoed by the President strikes a fair and reasonable balance between financial investment and environmental quality.... If we cannot swim in our lakes and rivers, if we cannot breathe the air God has given us, what other comforts can life offer us?[4]

But the exchange between Nixon and Congress suggested questions that remain unanswered today: After two hundred years of neglect, can the integrity of the nation's rivers, lakes, and coastal waters be restored? Is the nation willing to pay the short-term price for long-term improvements in the quality of our waters and aquatic resources? Are the benefits of clean water worth the cost? How successful has the Clean Water Act been in restoring our aquatic ecosystems? Do we even have the information needed to answer these questions? This study examines these questions two decades later.

Chapter 1 examines the state of U.S. waters before the 1972 Clean Water Act and reviews the basic goals and policies established by the law. Chapter 2 evaluates the best available evidence of how well the act has worked and looks at what more needs to be done to win the war against water pollution. This review suggests that, while much progress has been made in reducing chemical pollution, large amounts of toxic and other chemicals continue to be dumped into our nation's waters each day. Worse, while we are making some progress on the chemical front, we are moving backward in our efforts to restore the overall biological health of the nation's waters. Chapter 3 evaluates the economics of clean water by reviewing available information on the price we pay for this continuing pollution. Chapters 4 through 7 examine programs designed to protect the health of our waters and looks at how well they have worked in the real world. The last chapter suggests specific changes to the Clean Water Act that would help close the gap between the goals established in 1972 and the current state of our surface waters.

Chapter 1
The Need for Clean Water

The Impetus for the Clean Water Act

By the early 1970s, water pollution had reached crisis proportions in the United States. The most dramatic alarm rang on June 22, 1969, when the Cuyahoga River in Cleveland, Ohio, burst into flames, fueled by oil and other industrial wastes.[1]

But the Cuyahoga conflagration was not an isolated accident. In 1971, a task force launched by Ralph Nader issued a detailed report, *Water Wasteland*, outlining the serious state of U.S. waters. Much of the report was anecdotal, as little comprehensive information existed on the condition of U.S. waters. Yet some data were based on nationwide studies, and the pattern portrayed by Nader's researchers was telling:

- The Department of Health, Education and Welfare's Bureau of Water Hygiene reported in July 1970 that 30 percent of drinking water samples had chemicals exceeding the recommended Public Health Service limits. The Detroit River contained six times the Public Health Service limit for mercury.
- The Food and Drug Administration (FDA) reported in February 1971 that 87 percent of swordfish samples had mercury at levels that were unfit for human consumption.
- A national pesticide survey conducted in 1967–68 by the U.S. Bureau of Sport Fisheries (now part of the U.S. Fish and Wildlife Service [FWS]) measured DDT in 584 of 590 fish samples, with levels up to nine times the FDA limit. In 1969, the FDA seized 28,150 pounds of Lake Michigan coho salmon that had been contaminated.
- Indiana's Brandywine Creek was declared unfit for swimming in 1969. The Hudson River contained bacteria levels 170 times the safe limit.
- Record numbers of fish kills were reported in 1969. Over 41 million fish were killed, more than in 1966 through 1968 combined, including the largest recorded fish kill ever—26 million killed in Lake Thonotosassa, Florida, due to discharges from four food-processing plants.

5

- A 1966 survey found that almost 2 million acres of shellfishing beds had been closed due to pollution.
- Ecologist Dr. Charles F. Wurster, Jr., cautioned, in a recapitulation of Rachel Carson's historic warning in *Silent Spring* almost a decade earlier, that we could lose fifty to one hundred species of birds by the turn of the century due to toxic chemicals.
- A 1968 survey found that pollution of the Chesapeake Bay caused $3 million in losses to the fishing industry, and Federal Water Quality Administration economist Edwin Johnson estimated that water pollution cost the nation $12.8 billion a year.[2]

The Nader report, which thrust water pollution into the light of national media attention, was confirmed by official government sources. In 1971, the *Second Annual Report* of the President's Council on Environmental Quality (CEQ) paralleled, although in less dramatic language, many of the findings of *Water Wasteland*. More than one-fifth of the nation's shellfish beds were closed because of pollution, and the annual commercial harvest of shrimp from coastal areas had dropped from more than 6.3 million pounds before 1936 to only 10,000 pounds in 1965.[3] The number of fish reported killed each year from pollution increased from 6 million in 1960 to 15 million in 1968 and 41 million in 1969.[4] According to EPA estimates, almost one-third of U.S. waters were "characteristically polluted," that is, known to violate numeric federal water quality criteria. Less than 10 percent of U.S. watersheds were characterized by EPA as unpolluted or even moderately polluted.[5]

The Vision of the Clean Water Act

Prior to 1972, efforts had been made to address water pollution at both the federal and state levels.[6] The Water Pollution Control Act of 1948 (P.L. 80-845) provided the first federal funds for state water pollution control programs and the first dribble of subsidies for the construction of sewage treatment plants. All of the details, however, were left to the states. Uncle Sam began to subsidize sewage treatment construction more seriously with the Federal Water Pollution Control Act of 1956 (P.L. 84-660), with increasing commitments in 1961 (P.L. 87-88), 1965 (P.L. 89-234), and 1966 (P.L. 89-753). Yearly federal expenditures rose from $50 million in FY 1961 to $1.25 billion in FY 1971, and by 1972, more than thirteen thousand grants had been started in about ten thousand locations.[7]

For the most part, however, enforceable mandates or standards did not back up this increasing flow of federal dollars. No federal requirements were imposed on industrial polluters, and municipal dischargers benefited from federal dollars without any significant accompanying federal controls. Most notably, industries and cities did not need federal permits to discharge wastes into waterways. Enforcement was narrow and infrequent and generally limited to interstate pollution. The only serious effort came in 1965, when Congress created the Federal Water Pollution Control Administration (FWPCA) and required the states to develop water quality standards for interstate waters. Even then, enforcers had to prove that a particular polluter caused violations of these instream standards—no small task given the primitive state of water quality monitoring and science and the crowd of dischargers to most polluted waters.

These early efforts clearly were inadequate to cure the serious ills afflicting the rivers, lakes, and coastlines of the United States. By 1972, the nation was ready for stronger medicine. Senator Ed Muskie summarized the situation as he urged his colleagues to override President Richard Nixon's veto of the 1972 law:

> Our planet is beset with a cancer which threatens our very existence and which will not respond to the kind of treatment that has been prescribed in the past. The cancer of water pollution was engendered by our abuse of our lakes, streams, rivers and oceans; it has thrived on our half-hearted attempts to control it; and like any other disease, it can kill us.
>
> We have ignored this cancer for so long that the romance of environmental concern is already fading in the shadow of the grim realities of lakes, rivers and bays where all forms of life have been smothered by untreated wastes, and oceans which no longer provide us with food.[8]

Inspired by these concerns, Congress began the 1972 act with the underlying visions and goals that were absent from previous federal water pollution laws. Most important, Congress declared, "The objective of this Act is to restore and maintain the chemical, physical and biological integrity of the Nation's waters."[9] (This chapter sets forth only the most basic explanation of this extremely complex law.)

The language of this bedrock objective is significant in two critical respects. First, by defining the target in terms of ecosystem integrity, Congress sought not just to stop the bleeding, but to return the patient to full health—to rid our waters of all human impacts that threaten human health and the health of aquatic ecosystems. Second, by insisting that we restore *and* maintain our aquatic ecosystems, Congress directed not only that we repair damaged waters, but that we actively protect those waters that so far have escaped the impacts of past pollution, that is, that we keep clean waters clean.

Congress also set forth a number of subsidiary and interim goals: "In order to achieve this objective it is hereby declared that, consistent with the provisions of this Act—

(1) it is the national goal that the discharge of pollutants into the navigable waters be eliminated by 1985;

(2) it is the national goal that wherever attainable, an interim goal of water quality which provides for the protection and propagation of fish, shellfish, and wildlife and provides for recreation in and on the water be achieved by July 1, 1983;

(3) it is the national policy that the discharge of toxic pollutants in toxic amounts be prohibited.[10]

The first goal is commonly known as "zero discharge," the second as "fishable and swimmable waters," and the third as "no toxics in toxic amounts."

All of the efforts by EPA and the states to cleanse the nation's waters of pollution were supposed to be driven by these pronouncements, and any fair effort to judge the success of these programs must use them as the "guiding star."[11] Senator Muskie, at least, meant what he said: "These are not merely the pious declarations that Congress so often makes in passing its laws; on the contrary, this is literally a life or death proposition for the Nation."[12]

As important as the principal objective and the interim goals of the law was Congress's newfound fortitude in supporting theory with on-the-ground controls, spurred by the Senate finding that the prior approach "has been inadequate in every vital aspect."[13] The heart of the law is embodied in section 301, which instructs that no one has a right to use the nation's waters as dumping grounds for pollution: "Except as in compliance with [specific provisions of] this Act, the discharge of any pollutant by any person shall be unlawful."[14] Pollution would be allowed to continue only as required by the limitations of technology and economic achievability. A basic corollary to this principle was that pollution control was to be achieved by reducing pollutants, not by diluting them in receiving waters.[15] This change shifted the burden from the government, which no longer had to prove harm to justify action, to the discharger, which had to explain why the discharge could not be eliminated.

The U.S. Army Corps of Engineers (the Corps, or COE) issued some water pollution control permits under the antiquated 1899 Rivers and Harbors Act, but not in a systematic—and certainly not in a universal—fashion. The essential progress made in 1972 was to require permits for all point sources of pollution and to define tough new requirements for these permits. Municipal sewage dischargers had to provide at least secondary treatment,[16] and industrial dischargers had to meet analogous minimum control levels defined by EPA on the basis of what the "best technology" could achieve. These new baseline obligations, known as "technology-based controls," were intended to achieve across-the-board pollution reduction (and, wherever possible, elimination) and to create a level playing field for most dischargers. An improved version of the water-quality–based controls of the 1965 law remained as an important backstop, however, and all dischargers were required to achieve stricter controls where necessary to assure that water quality standards were met in specific waters. States were required to develop water quality standards for in-state as well as for interstate waters, to identify all waters not meeting these standards, to calculate the additional pollution reductions needed to achieve the standards, and to incorporate these requirements into permits. If a state failed to perform any of these functions, EPA was required to do so instead.[17]

These strict new requirements for discrete dischargers were designed to address some of the goals of the 1972 law—to have "fishable and swimmable waters" by 1983, "zero discharge" by 1985, and "no toxics in toxic amounts." But even in 1972, Congress recognized that much of the country's water pollution came from far more diffuse, *nonpoint* sources[18] (polluted runoff from farms, streets, parking lots, mining sites, and so on) and that the ultimate objective of the act, restoring the "chemical, physical and biological integrity of the Nation's waters," required a more comprehensive approach. Even the definition of *pollution* in the 1972 law, "the man-made or man-induced alteration of the chemical, physical, biological, and radiological integrity of water,"[19] suggested impacts far broader than the release of chemical pollutants from sewers and factories. In section 208 of the new law, Congress urged on EPA and the states a system of comprehensive water quality planning and management. States were to identify all sources of pollution and water-body impairment and to develop a coordinated approach to address these forms of pollution simultaneously. The watershed approach, largely abandoned by EPA and the states in the 1980s, resurfaced as a prominent "new" approach to pollution control as the act passed its twentieth anniversary.

The Clean Water Act has been modified extensively since 1972, most notably in 1977[20] and 1987.[21] In 1977, for example, Congress expanded and specified EPA's mandate to control the release of toxic pollutants into sewers and surface waters. In 1987, frustrated by slow progress in controlling pollution from diffuse sources, Congress adopted new programs to address polluted runoff from farms, factories, and city streets. In an effort to return to more site-specific, watershed-based planning, it adopted special programs to clean up the Great Lakes (by improving and implementing the 1978 International Great Lakes Water Quality Agreement), the Chesapeake Bay, and seriously impaired estuaries around the country.

None of these changes, however, altered the fundamental objectives of the act to eliminate all forms of pollution and to restore and maintain the integrity of the nation's waters.

The Need to Meet the 1972 Goals

In its fifth annual survey of the best places to live in the United States, *Money* magazine ranked what the people of this country most want in a city. The highest valued characteristic, even above low crime rate, was clean water. The local leisure activity with the highest rating was access to a lake or an ocean.[22] Both measures indicate the high value placed on the quality of our water resources.

More rigorous surveys underscore the public's concern about water pollution. In a 1992 Roper poll, 77 percent of the respondents agreed that water pollution was one of the "most serious" environmental problems (see Table 1.1). An even higher percentage (79) believed that current water pollution regulations do not go far enough to protect public resources.[23]

A comprehensive review of more than five hundred public opinion surveys conducted since 1974[24] confirms that the public ranks water quality high among environmental problems (but rejects *Money*'s conclusion that the public views water quality problems as more serious than such societal ills as crime and drugs). According to this survey, most people believe that water quality problems are getting worse, and the percentage of people who share this view has increased since the Clean Water Act was passed.[25]

These responses indicate that the U.S. public understands, at a fundamental level, the importance of clean water and healthy aquatic ecosystems to their lives and welfare. And for good reasons:

● The human body is more than two-thirds water; to replenish this supply, we each consume an average of 2 liters of fresh water each day (directly or through other liquids). Water is the most basic component of all of our cells, the supply system for nutrients (through our blood), the garbage disposal for many of our wastes (through urine and perspiration), and a structural and mechanical mainstay for most of our vital organs.

● Water is essential to all of our food supplies—to irrigate our crops, to water our livestock, and to spawn and rear our fish and shellfish.

● Since all life on earth evolved from the oceans, water fuels all species—plant and animal—and is the most fundamental component of all of our ecosystems, whether we think of them as marine, freshwater, or terrestrial.

● Throughout history, water bodies have supported human cultural develop-

TABLE 1.1

Public Perceptions on Water Quality Issues:

PERCEIVED SERIOUSNESS OF ENVIRONMENTAL PROBLEMS

	NOT THAT SERIOUS	FAIRLY SERIOUS	MOST SERIOUS
Water pollution	3%	19%	77%
Toxic waste sites	3	23	72
Drinking water shortage	8	21	68
Air pollution	7	28	64
Ozone layer	8	24	62
Landfill shortage	8	32	58
Loss of natural areas	13	39	46
Global warming	14	29	45
Endangered species	17	34	45
Loss of wetlands	14	35	42

HOW FAR DO ENVIRONMENTAL LAWS AND REGULATIONS GO?

Not far enough	63%
Just right	17
Too far	10
Don't know	10

OPINIONS OF CURRENT U.S. LAWS AND REGULATIONS

	GONE TOO FAR	STRUCK RIGHT BALANCE	NOT FAR ENOUGH
Water pollution	3%	13%	79%
Air pollution	5	18	72
Natural areas	7	27	59
Endangered species	8	24	53
Wetlands	11	31	51

Source: The Roper Organization, *Natural Resource Conservation: Where Environmentalism Is Headed in the 1990s,* The Times Mirror Magazines National Environmental Forum Survey (June 1992).

ment, serving as sources of commerce, recreation, and aesthetic and spiritual fulfillment.

Clean water and healthy aquatic ecosystems can provide all of these vital functions and more. Pollution and crippling of our rivers, lakes, and coastal waters, on the other hand, can have devastating effects on our society and our economy. Contaminated drinking water, beaches, and seafood can cause serious illness and even death. Lost or damaged habitat can lead to extinctions or serious declines in species that live in, or depend on, aquatic ecosystems. Chemical pollution can kill large numbers of fish and wildlife, and toxic chemicals, in even small amounts, can cause deformities, reproductive failures, and other adverse effects on many species, including humans. Polluted waters and lost habitat can also severely affect important sectors of our economy—commercial and recreational fishing and shellfishing, water-based recreation and tourism, and waterside land among them.

Less tangible but equally important is the value of clean water and healthy ecosystems to our spiritual well-being. Loren Eiseley wrote, "If there is magic on this planet, it is contained in water."[26] From early in our religious and cultural history, water has been a symbol of purity and renewal. It has been a source of pleasure and relaxation—from swimming to fishing to simply strolling along the beach—throughout history. For these reasons, as well as for healthy food and drinking water, people have built their communities along the shores of water bodies. Ancient mariners rowed and sailed away from the safety of the shore not only for early commerce, but because they were attracted by the beauty and majesty of these great water bodies and the creatures that inhabited them. European colonists who explored and settled North America relied heavily on surface waters as a natural highway system. Today, millions flock to pristine mountain streams, placid blue lakes, and sandy white beaches for fun and leisure and for the pure wonder of gazing over clean, healthy waters. From the rocky shores of Maine to Florida's Everglades, from the mountain pools of Colorado to the Great Lakes of the Midwest, from the trout streams of Montana to the mighty Mississippi, and from the magnificent canyons of the Colorado to the sandy beaches of California, the people of this country continue to spend huge amounts of time in or near the water.

We will explore these values and issues in detail as we review the successes and failures of the Clean Water Act over the past two decades. The most fundamental questions to ask are not how many standards have been issued and how many permits have been written, but how successful has the law been in protecting people and aquatic ecosystems? Chapter 2 focuses on these questions.

Chapter 2
The State of Our Waters
Twenty Years Later

How can we measure the success of a statute as complex as the Clean Water Act? The law is implemented by EPA, several interstate agencies, fifty states (and several territories), and thousands of local governments. Its goal is to restore and maintain more than 3 million miles of rivers and streams, almost 27 million acres of lakes, and more than 35,000 square miles of estuaries.[1] Some waters are monitored frequently, others less often, and many not at all. Even the best monitoring efforts vary in frequency, methods, quality, and a host of other ways. And all of these factors have changed considerably over the two decades since the law was passed.

Should success be judged by numbers of treatment systems built or pounds of pollutants reduced? By the number of waters that meet water quality standards, or by percent improvement or degradation in chemical water quality measures? In section 1 of this chapter we look to these traditional measures of water quality as EPA and the states report them to Congress.

But does the public (or Congress) really care about such abstractions, and do these numbers really tell us what we want to know? Instead of relying on traditional measures, we searched for answers to the practical questions that most U.S. citizens have about their water resources. In section 2 we explore how well the Clean Water Act has served to protect public health: Can I drink my tap water? Can my family eat fish and shellfish without fear of becoming ill? Can my children swim at their favorite lake or beach? In section 3 we ask similar questions about the ecological health of our waters: Are fish and wildlife species getting healthier, or are populations plummeting? How many wetlands have been filled? How many floodplains have been developed? How many miles of streamside habitat have been lost?

Taken together, these measures present a far more complete and balanced picture of how well the Clean Water Act has protected human health and aquatic ecosystems over the past twenty years. Particularly interesting are comparisons of the traditional measures used by agency officials with real-world indicators of environmental quality. For example, how well does a state's assessment of whether its waters support healthy aquatic populations match up with the numbers of threatened and endangered species in that state? Do state assessments of "fishable waters" square with information on shellfish bed closures or fish kills? In many cases, the inconsistencies are more informative than the parallels.

Traditional Measures of Pollution and Water Quality

Improvements in Pollution Control Technology

The first traditional way to evaluate success in water pollution control is to count the numbers of new treatment systems installed and the pounds of pollutants removed by those systems. By this measure, the United States has made significant progress since (and even before) passage of the 1972 Clean Water Act, but it still has a long way to go to eliminate water pollution altogether.

The federal government invested $56 billion in municipal sewage treatment from 1972 to 1989,[2] with total federal, state, and local expenditures of more than $128 billion. These investments gained impressive results. The percentage of the U.S. population served by wastewater treatment plants jumped from 42 in 1970 to 67 in 1975, to 70 by 1980, and up to 74 by 1985.[3] As of 1988, plants providing secondary treatment or better served 58 percent of the U.S. population.[4] This improved treatment, according to EPA, has reduced annual releases of organic wastes by 46 percent, despite a large increase in the amount of wastes treated.

The same measure viewed from the opposite direction, however, shows a glass only half full. By 1988, public sewer systems serving 26.5 million people in the United States provided less than secondary treatment, and 1.5 million people had no treatment at all, with raw sewage discharged into public waters.[5] About 70 million people were not served by public sewers at all;[6] while many of these people have properly designed and maintained septic systems, others have in-ground systems that leak pollutants into surface or groundwater.[7] In 1990, EPA estimated that the cost of meeting these additional municipal treatment needs through the year 2010 would exceed $110 billion (in 1990 dollars).[8] Judged by these investment needs, while our municipal pollution control efforts have taken a giant step since 1972, we are still only halfway to our destination. Some of these needs are for advanced treatment systems to reduce discharges of nutrients; others are to tackle ongoing releases of raw sewage into the nation's waters. A 1992 NRDC study (see Table 2.1),[9] for example, showed that fourteen large cities with combined sewer systems had deposited into surface waters more than 165 billion gallons of raw sewage mixed with

TABLE 2.1

Estimated Annual Combined Sewer Overflow Releases from 14 U.S. Cities

CITY	FLOW (BILLION GAL)	SEDIMENTS (MILLION LB)	ORGANIC WASTES (MILLION LB)	COPPER (LB)	LEAD (LB)	ZINC (LB)
Atlanta	5.3	15.0	5.5	4,500	15,000	15,000
Boston	5.2	9.4	4.3	3,900	7,900	11,000
Bridgeport	1.7	1.1	0.4	1,500	4,900	5,000
Chicago	27.0	10.0	6.9	21,000	4,400	144,000
Cleveland	5.9	26.0	4.7	6,700	9,200	12,000
Minneapolis/ St. Paul	1.6	2.5	0.7	800	1,500	3,500
Narragansett	2.6	3.5	2.3	3,000	1,700	7,000
New Bedford	1.1	1.7	0.5	1,000	3,300	3,300
New York	84.0	83.0	38.0	71,000	240,000	240,000
Philadelphia	20.0	23.0	17.0	17,000	58,000	58,000
Richmond	4.1	9.1	2.5	3,500	12,000	12,000
San Francisco	1.7	1.8	1.5	1,500	4,900	5,000
Seattle	2.9	2.8	1.5	2,100	4,200	5,400
Washington, D.C.	2.2	5.4	0.9	1,900	5,500	5,200
TOTAL	165.3	194.3	86.7	223,200	372,500	526,400

Source: NRDC, When It Rains . . . It Pollutes (April 1992).

polluted storm water and industrial discharges each year. About eleven hundred such systems exist nationwide, and we estimate that combined sewer overflows (CSOs) nationwide release between 3 billion and 11 billion pounds of solids and between 1 billion and 3 billion pounds of organic matter each year.[10]

Similar gains are evident in the industrial sector. In 1973, industry spent about $1.8 billion on water pollution controls (including both capital and operating and management—O & M expenses). By 1986, this had jumped to almost $5.9 billion.[11] But *new* water pollution equipment expenditures seem to have peaked in the late 1970s, reflecting a substantial curb in EPA's issuance of new industrial water pollution controls. Total industrial pollution control expenditures over this period exceed $57 billion (see Table 2.2).

Again, these investments have reaped large dividends in total pollution reductions. According to EPA, pollution controls implemented in twenty-two industries since 1972—under a Consent Decree between EPA and NRDC— have reduced releases of selected "priority" toxic organic pollutants by 99 percent, or by almost 660,000 pounds per day. (In 1976, EPA developed a list of 129 priority toxic pollutants on which it would focus its toxic water pollution control efforts; three pollutants have been removed from this list.) Reductions in toxic metals are estimated at almost 98 percent, or more than 1.6 million pounds per day.[12] All told, assuming EPA's estimates are correct, these controls have eliminated the release of more than 1 billion pounds of toxic pollutants each year into the nation's rivers, lakes, and coastal waters. Even higher amounts of conventional pollutants, like organic wastes and solids, have been controlled with this new technology.[13]

As with sewage treatment plants, however, the picture looks different in light of how much pollution industries continued to release in 1992. According to EPA's most recent *Toxics Release Inventory* (*TRI*), in 1990, U.S. industries reported releases of almost 200 million pounds of toxics into surface waters, more than 2.5 million pounds of which were carcinogens. (The *TRI* is a compilation of industry-generated reports of releases of toxic chemicals into the environment under the Emergency Planning and Community Right-to-Know Act of 1986.) Almost another 450 million pounds a year were released into public sewers.[14]

Even these numbers are far from complete. The *TRI* reporting system does not include hundreds of toxic chemicals and dozens of types of facilities that generate toxics. It covers only about three hundred chemicals and only manufacturing industries, and it exempts facilities that manufacture or process less than 25,000 pounds of those chemicals each year.[15] It does not cover hazardous waste treaters, for example, which by themselves release more than 300 million pounds of toxics each year, including more than two hundred toxic pollutants, at least sixty of which may cause cancer and at least twenty-three of which may cause birth or genetic defects.[16] Thus, while *TRI* data show large continuing releases of toxic chemicals into our surface waters, these numbers are likely to be conservative.

TABLE 2.2

Industry Expenditures on Water Pollution Control (Billions of 1986 $)

YEAR	CAPITAL	O & M	TOTAL
1973	1.280	1.497	1.821
1974	1.009	1.261	2.270
1975	1.280	1.497	2.777
1976	1.599	1.824	3.423
1977	1.674	2.203	3.877
1978	1.263	2.550	3.813
1979	1.262	3.040	4.302
1980	1.163	3.220	4.383
1981	1.028	3.554	4.582
1982	0.977	3.489	4.466
1983	0.819	3.943	4.762
1984	0.888	4.296	5.184
1985	1.018	4.610	5.628
1986	1.039	4.820	5.859

Source: Bureau of the Census, MA-200(73)-1-(86)-1.

Still, despite loopholes and limitations in the reporting system, and assuming the accuracy of industry reporting, the four years of *TRI* data (1987–90) show continuing declines in the release of toxic chemicals into our waters and sewers (see Table 2.3).

All in all, these measures show considerable progress in reducing the total amounts of pollution reaching U.S. surface waters from specific sources. Equally clearly, however, we are still a long way from eliminating the discharge of pollutants into our waters, a goal Congress set for 1985. And while EPA agrees that reductions in point source pollutant releases are one useful mea-

TABLE 2.3

Industrial Toxic Pollutant Trends (Millions of Pounds Released per Year)

	1987	1988	1989	1990
Surface	412	311	193	197
POTWs*	610	574	557	447

Source: EPA, *1990 Toxics Release Inventory,* Public Data Release (May 1992), 62.
Note: Data are estimates from manufacturers subject to TRI reporting.
*Publicly owned treatment works.

sure of progress and the effectiveness of the regulatory program, it acknowledges that the availability and quality of data are limited, especially for toxics.[17]

Long-Term Water Quality Trends

A second traditional measure of water quality success is whether levels of pollutants in the water have diminished, increased, or remained constant. Surprisingly little information is available to answer this question. Where information is present, it shows that some water bodies have improved due to reduced pollution from factories and sewage, while others have degraded due to increased runoff from agriculture and other sources. Because of the predominance of pollution from runoff and air deposition, however, most streams show no significant trends.

Thousands of water quality monitoring stations exist around the country, but relatively little of the information collected at these stations is suitable to determine long-term water quality trends. Monitoring and analytical methods, water quality conditions measured, and consistency of data vary significantly. Some stations have been in place for long periods of time, others for only short durations. These problems led one commenter to refer to water quality knowledge in this country as "data-rich but information-poor."[18]

The one major exception is the National Stream Quality Accounting Network (NASQAN) run by the U.S. Geological Survey (GS), which consists of 403 stations nationwide located at GS stream-gauging stations.[19] Because a single federal agency runs this program, it can assess national trends without variations among state programs. The number and location of monitoring stations and the scope of pollutants measured, however, are necessarily limited.

In data from 1978 to 1987, the majority of NASQAN stations showed no significant trends, that is, no significant improvement or deterioration. This probably reflects the fact that most of the stations are relatively removed from major point sources (factories or sewage plants). Thus, the data are more likely to reflect water quality effects from land use and atmospheric deposition than from discrete sources and are not likely to reflect the reductions in industrial or sewage pollution discussed in the previous section. This does not mean that point source reductions are insignificant; it simply means that the water quality improvements achieved over the past twenty years are more localized than can be measured by 403 stations distributed across the country.

Definite trends were detected, however, in concentrations of some pollutants in some waters.

● Nitrogen and Dissolved Solids. Four times as many streams deteriorated as improved. This trend is consistent with increased polluted runoff from agriculture.

● Phosphorus. Of the streams with clear trends, 85 percent showed improvements. This may reflect reductions from some point sources, such as sewage treatment plants.[20]

● Dissolved Oxygen, Bacteria, and Metals. More streams improved than degraded, again reflecting improvements in point source controls.

● Some Metals (Lead, Zinc, Chromium, Silver, and Arsenic). Only a handful of streams degraded; many more improved. In the case of lead, these reductions can be explained by the elimination of lead in gasoline, which formerly reached waterways through runoff from roads.[21]

A more recent GS analysis of water quality trends for conventional pollutants from 1980 to 1989 generally confirmed these conclusions. Again, the vast majority of stations showed no trend. Where trends did occur, more streams showed improvement than degradation for all pollutants except dissolved solids, again consistent with increasing pollution from agricultural and other sources of runoff.[22] A number of streams still reflected worsening conditions, however, ranging from one out of 306 streams measured for chromium to 115 of 385 streams measured for acidity and alkalinity, confirming that overall water quality progress is inconsistent.

More definitive information is available on specific waterways, and dramatic improvements are evident in some of the severe forms of pollution that persisted before 1972. Of course, it is difficult to draw nationwide conclusions from a small sampling of waters. Nevertheless, a recent analysis of water quality progress in three estuaries came to conclusions similar to those reached through an evaluation of the nationwide NASQAN data. In a study of the

Delaware River estuary, the Flint River, Georgia, and the Neches River, Texas, Dr. Ruth Patrick of the Academy of Natural Sciences in Philadelphia found the following:

1. Even for specific water bodies, the absence of consistent monitoring data over time renders evaluation of water quality trends difficult.

2. Significant improvements can be detected in water quality conditions in some parts of some of the waters studied (or for some types of pollutants), in particular due to reduced pollution from factories and sewage treatment plants.

3. Nevertheless, water quality has degraded in other portions of the waters studied (or for other types of pollutants). In some cases, these problems resulted from persistent or increased discharges from factories or from continued discharges of sewage from CSOs. In other cases, continued or worsening problems resulted from polluted runoff from new development, agriculture, and other sources.[23]

Other major water systems, especially the Great Lakes and the Chesapeake Bay, have benefited from concerted multijurisdictional assessment and cleanup efforts. Because of the magnitude and duration of these programs, more information is available about long-term trends for them than for many other water bodies. Not surprisingly, some areas of these ecosystems show major improvements, but significant problems remain in both the Great Lakes and the Chesapeake Bay.

In the 1960s, the Great Lakes became a major symbol of water pollution when some scientists pronounced Lake Erie "dead." Severe levels of phosphorus caused large blooms of algae (*eutrophication*), which robbed the water of oxygen and caused tremendous fish kills.[24] A massive cleanup effort on both sides of the U.S.–Canadian border reduced phosphorus levels substantially, largely through the construction or improvement of sewage treatment plants.[25] Even with these reductions, however, phosphorus levels have not dropped to target levels set jointly by the United States and Canada, and oxygen depletion remains a concern in Lake Erie and thirty other sites in the region.[26] Meanwhile, concentrations of nitrates and nitrites, which also can lead to eutrophication, have increased over the past two decades, largely due to runoff from agriculture.[27]

Increased evidence of severe toxic contamination in Great Lakes water, fish, and sediment—at levels that the U.S. General Accounting Office (GAO) concluded "raise serious questions about [the] future [of the Lakes]"[28]—has rallied public interest in a new round of cleanup. The GAO cited harm to many species of fish and wildlife and serious human health threats, from developmental problems in children to increased risk of cancer, at even low levels of exposure.[29] In its *Sixth Biennial Report on Great Lakes Water Quality,* the International Joint Commission on the Great Lakes (IJC) referred to the presence of persistent toxic substances as a "daunting, unresolved challenge."[30] The IJC noted that, while the Great Lakes Agreement "calls for the virtual elimination" of the discharge of persistent toxic pollutants, "[w]e have not yet virtually eliminated, nor achieved zero discharge of any persistent toxic substance."[31]

A similar picture is evident in the Chesapeake Bay, where a concerted cleanup effort has reaped benefits over the past decade, but where serious water quality problems remain. The Chesapeake Executive Council (the governing body for the Bay cleanup), for example, reports that, while phosphorus levels have declined by 20 percent since 1985, total nitrogen loadings increased from about 120,000 pounds per day in 1972 to more than 160,000 pounds per day in 1989.[32] A 1991 book written for the Chesapeake Bay Foundation cites a wide range of continuing water pollution problems in the Bay. Water area with oxygen depletion rose fifteen-fold from 1950 to 1980; at times during the 1980s, as much as 40 percent of the Bay's water was below healthy oxygen levels.[33] And as in the Great Lakes, toxic pollutants are of increasing concern throughout much of the Chesapeake Bay: "Toxic chemicals . . . have been documented throughout the bay's waters and its sediments. Fish, shellfish, and crabs have been found to contain a range of toxics in Maryland and Virginia's bays and rivers during the last fifteen years."[34]

While most other major water bodies do not enjoy coordinated assessment and cleanup efforts similar to those in the Great Lakes and Chesapeake Bay, recent assessments of major rivers show a similar checkered pattern of successes and failures over the past twenty years:

Mississippi River. An effort to assess water quality in the Mississippi was plagued by an absence of data, variable or unknown data quality, and limited observation sites and times. Nevertheless, the analysis documented serious water quality problems from sedimentation, nitrogen and phosphorus, pesticides, and other toxic pollutants.[35]

Columbia River. The construction of a large system of dams has altered temperature and dissolved gas, oxygen, and salinity levels in much of the Columbia River and its tributaries. While some of these impacts have been mitigated, problems remain. Nutrient levels have declined in the basin due to sewage treatment plant construction and improvement, but some eutrophication continues. Suspended sediments have increased due to erosion from logging and farming, but, ironically, the presence of dams offsets this effect. Data on the presence of toxic pollutants is severely limited, but available information indicates the presence of pesticides, PCBs, and metals such as cadmium and lead.[36]

Missouri River. As in many water bodies, high levels of bacteria and other pathogens, oxygen depletion, and other gross forms of pollution dominated water quality concerns in the Missouri in the 1960s and 1970s. As improved sewage treatment began to address these impacts, the focus turned to toxic pollutants, including pesticides from agriculture, and heavy metals and organic pollutants from mining and industry. While cleanup efforts have reduced some sources of toxic pollutants, evidence shows that serious toxic pollution continues in the watershed.[37]

Ohio River. Relatively good, long-term water quality data are available since at least 1949 through the efforts of the Ohio River Valley Sanitation Commission (ORSANCO). Through much of this century, water quality in the Ohio River declined dramatically, with evidence of problems related to turbidity, solids, salts, nitrates, oxygen depletion, bacteria, and toxics. Turbidity decreased during the 1950s through the 1980s, probably due more to flood control than to pollution control. Construction of sewage treatment facilities during the 1970s improved water quality for dissolved oxygen, bacteria, and other pollutants, but oxygen levels remain low below cities on the main stem of the Ohio, and bacteria levels remain high, particularly in upper portions of the river. As for other waters, recent data have raised concerns about toxic pollutants, with frequent violations of water quality standards for some metals and with increased levels of arsenic and lead. While the detection of organic pollutants has diminished, ORSANCO continues to issue fish consumption advisories due to the presence of persistent toxic pollutants in fish.[38]

All of this information, taken together, confirms that progress has been made in reducing the release of some pollutants by some sources. These reductions have improved overall water quality in some waters, often dramatically. Still, many other waters continue to be polluted by other sources, such as runoff from logging, farming, mining, and urbanization. More traditional sources of pollution, such as factories and sewage treatment plants, continue to release toxic and other pollutants in significant amounts. Overall, long-term progress continues to be difficult to assess due to serious shortages of data and problems with data consistency and quality. Clearly, we must evaluate other measures to determine how well we have met the overall goals of the Clean Water Act.

Attainment of Designated Uses

The third traditional way to determine how well the Clean Water Act is working, and the one with the clearest legal imprimatur, is the system of biennial water quality reports required by the Clean Water Act. These reports continue to show that a large percentage of the nation's waters remain unsafe for fishing, swimming, and other uses. But serious problems with the reporting system render this information incomplete and inconsistent, and the number of waters that remain polluted may be significantly understated.

Section 305(b) of the Clean Water Act requires states to submit to EPA every two years an analysis of water quality, including an assessment of which waters meet water quality standards. These reports must evaluate, among other things, the extent to which state waters meet the basic goals of the act. In turn, EPA is required to analyze these reports and submit a comprehensive analysis to Congress every two years.[39] These reports are known as the *National Water Quality Inventory*. The first inventory was released in 1974; the most recent was issued in April 1992 and covers the years 1988–89.[40]

The most fundamental tools used by EPA and the states to prepare these reports are water quality standards. These standards consist of two essential components—designated uses and criteria designed to protect those uses. Designated uses are simply the public uses and values for which the waters are to be protected, such as fishing, swimming, drinking, protecting fish and wildlife, and so on. While states have discretion to determine appropriate uses for their waters, the act requires that they protect the "fishable and swimmable" uses at a minimum, wherever these are attainable. Water quality criteria are specific chemical limits or other measures deemed necessary to protect those designated uses. Again, each state determines appropriate water quality criteria, subject to review and approval by EPA.

The most recent *National Water Quality Inventory* demonstrates that even the interim goals of the 1972 Clean Water Act have not been met. According to the *Inventory*, at least a third of our rivers, half of our estuaries, and more than half of our lakes are not meeting or fully supporting designated uses. (Other reports, to be discussed later, suggest that these numbers are seriously understated.) Clearly we have not yet met the 1983 "fishable and swimmable" goal of the law. And despite incomplete monitoring, states report that toxic pollutants affect a large percentage of waters.[41] Thus, we have not eliminated the release of "toxic pollutants in toxic amounts" either.

But does this mean that water quality is getting better or worse? It would seem relatively simple to evaluate how well the act has worked by comparing the information in each of EPA's inventories to determine how many waters have improved or degraded. In fact, this analysis is complicated by a host of problems, including significant changes in the amount of data available over the years and in procedures used to analyze and present the available information.

The early 305(b) reports were extremely cursory and were based on information on a small percentage of waters. The 1974 report, for example, attempted to characterize water quality in just twenty-two major waterways.[42] In general, few monitoring data were available even for these waters, and water quality criteria existed for only a handful of pollutants. The *1990 Inventory*, by contrast, is based on detailed information from fifty-one states and territories and thousands of monitoring stations around the country. The report includes summaries of compliance with water quality standards in each state, broken down by type of water, reasons for impairment, and other information. Clearly, one cannot simply compare the 1974 report with the 1990 report and draw conclusions about how much progress has been made since the Clean Water Act was passed.

Even the *1990 Inventory*, however, is hardly comprehensive. States reported that only 53 percent of river miles, 69 percent of lake acres, and 75 percent of estuarine area were "assessed" for the report.[43] And even these claims are misleading, since "assessed" does not mean actually "monitored" for toxic and other pollutants or evaluated with rigorous biological surveys. Many waters are judged by more subjective information, such as incidental reports or visual surveillance. The 1990 report was based on actual chemical measurements for less than one-fifth of our rivers, streams, and lakes and about one-quarter of our estuaries.[44]

Despite these limitations, it is possible to evaluate some trends by comparing the last several *Water Quality Inventories*, which by now are issued in a relatively standard format, using relatively consistent terminology (see Figure 2.1A–2.1C).

This analysis produces surprising results. Given the reductions in total amounts of pollutants reaching our waters and the general declines in or absence of measurable trends in levels of ambient water quality, as shown in the previous sections, overall attainment of designated uses should be improving. Instead, considerably fewer waters are reported as meeting designated uses in 1988–89 than in 1982–83, for all three types of waters.

There are several possible explanations for this trend. First, the total percentage of waters assessed and the number of states reporting has increased steadily over the past decade. Water pollution may not be getting significantly worse; our knowledge of the magnitude of the problem may just be improving. Second, the standards by which we measure use attainment may be improving, or getting stricter. In fact, EPA is making progress in expanding the criteria by which use impairment is judged, leading some states to report higher levels of impairment than in previous years.[45] Again, pollution is not necessarily getting worse, but we do have a better understanding of the problem. Finally, the traditional measures of pollution reduction and water quality used to define suc-

FIGURE 2.1A

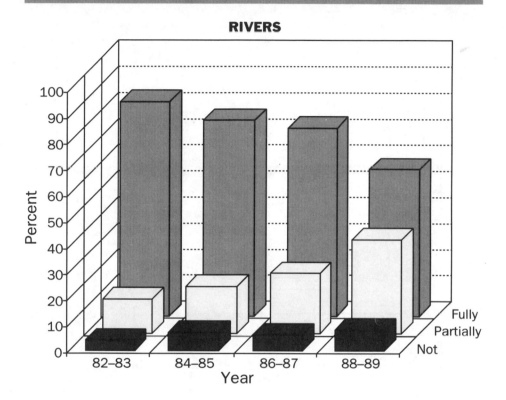

Trends in Water Quality Impairment
(% of Assessed Waters Fully, Partially, or Not Supporting Uses)

RIVERS

cess may be masked by other types of effects on aquatic ecosystems, such as degradation or elimination of aquatic habitat or concentration of pollutants over time even where current releases have been reduced—effects that will be discussed in more detail following.

The Environmental Protection Agency openly acknowledges that serious problems with the 305(b) reporting system lead to severe inconsistencies in how different states report use impairment:

> Unfortunately, the current value of the 305(b) reports as a source of environmental indicator data is severely limited. There are very large incon-

FIGURE 2.1B

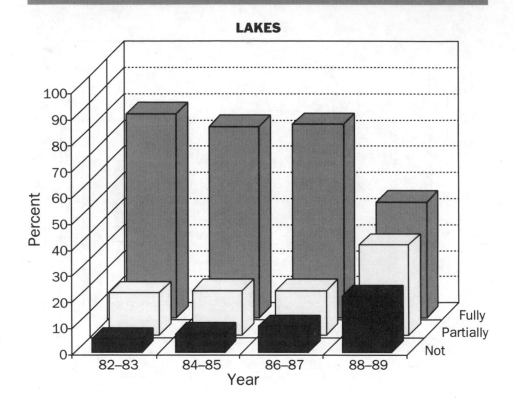

Trends in Water Quality Impairment
(% of Assessed Waters Fully, Partially, or Not Supporting Uses)

LAKES

sistencies among States in how water quality data are generated, analyzed and reported. States assess different subsets of their waters from one year to the next. In some instances, States even change their accounting of total waters from one year to the next. One problem in using this information for national reporting purposes stems from the considerable discretion that States have under the law in developing their own water quality standards.... As a result...making comparisons between States or trying to assess national status and trends is essentially impossible. And the inconsistencies in sampling design from year to year make it difficult

FIGURE 2.1C

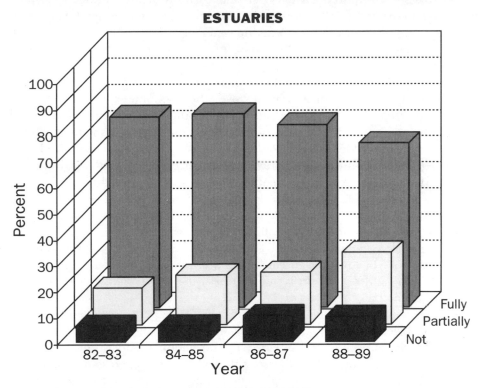

**Trends in Water Quality Impairment
(% of Assessed Waters Fully, Partially, or Not Supporting Uses)**

ESTUARIES

Source: EPA, *National Water Quality Inventories* (1984, 1986, 1988, 1990).

to assess trends even within individual States.[46]

While EPA is in the process of trying to inject more consistency into the process, even the agency's recommended criteria for 305(b) reporting encourage underestimates of use impairment by most if not all states. For example, under EPA's criteria for use impairment, states can conclude the following:[47]

● Uses are fully supported even if water quality criteria are exceeded in one out of ten samples. Uses are partially supported even with violations up to one-quarter of the time.

27

- Uses are fully supported if averages of samples are below criteria designed to protect drinking water. This means uses are supported even if nearly half of the samples violate drinking water protection criteria.
- A drinking water use is partially supported even if a drinking water advisory is in place for up to thirty days per year or if problems do not require closures, even if they affect treatment costs and quality of water (taste, odor, color, turbidity, and so on). Drinking water uses are "not supported" only if advisories last more than thirty days per year or if the system is actually closed.
- Uses are fully supported even if criteria to protect swimmers are exceeded in up to one out of ten measurements over the year. Since the criteria do not specify when these violations can occur, in theory, violations could occur much more often during peak swimming season, as long as water quality is better during colder months. Recreational uses are partially supported even if bathing areas are closed an average of once per year for less than one week. Thus, waters can be deemed "swimmable" even though they are known to pose risks to bathers.
- Uses are partially supported even if a "restricted consumption" fish advisory or ban is in effect for the general population or for sensitive subpopulations (pregnant women or children). That is, waters are labeled "fishable" even in the face of fish advisories or bans. Uses are not supported only if a "no consumption" advisory or ban is in effect or if there is a commercial fishing or shellfishing ban.

An even more fundamental question is what these reports suggest about the *ultimate* objective of the Clean Water Act—to restore and maintain the chemical, physical, and biological integrity of the nation's waters. The traditional measures of water quality—reductions in pollutant discharges and changes in water chemistry—are only surrogate indicators of ecosystem health. Even use attainment historically has been based primarily on chemical monitoring efforts. In the meantime, we must search for other indications of how well the Clean Water Act has worked to protect both human health and aquatic ecosystems.

To address some of these problems, EPA and many states are moving toward the use of biological water quality criteria, which compare the presence and overall health of a representative range of species in the water body with species in control waters that remain in a relatively natural state. First developed in 1981 by Dr. James Karr as the Index of Biotic Integrity, this technique and later variations of it rely on direct assessment of the populations, diversity, and community structure of fish, invertebrates, and other groups of species within an aquatic ecosystem.[48] The Environmental Protection Agency agrees that "[t]his indicator is the most direct measure possible of support of a Clean Water Act (CWA) goal, because maintaining biological integrity is one of the legislative mandates."[49]

Like all water quality criteria, biocriteria have some limitations that must be taken into account. One commenter has noted,[50] for example, that biocriteria take snapshots of ecosystem health, rather than examine the "ability of those systems to resist, recover from or otherwise respond to stress." Clearly, biocriteria do not measure the secondary effects of chemical pollution (such as long-term accumulation or threats to humans or wildlife that consume contaminated species). And current biocriteria detect impairment after the fact, rather than predicting future harm.

To address these problems, biological criteria must be used to supplement rather than to supplant other types of water quality criteria: In 1990, EPA issued national guidance indicating that use impairment should be based on any one of three types of data—chemical, biological, or toxicological impairment (regardless of findings of other data).[51] "Recent studies indicate that integrated assessments, including biological, chemical and physical data, convey a much more accurate and complete picture of water resource status than an approach based solely on chemical criteria."[52]

While some states have incorporated biological monitoring into regular networks, most use the procedures for only a relatively small percentage of waters.[53] It will be several years before biological criteria will generate more complete data on the full range of use impairment in the nation's waters. In the meantime, we must search for other indications of how well the Clean Water Act has worked to protect both human health and aquatic ecosystems.

Human Health Still Threatened

The Clean Water Act was enacted in part to address serious concerns about public health—polluted drinking water, contaminated fish and shellfish, and swimming beaches laden with bacteria. The first question, then, is how well has the act succeeded in protecting the public from these basic threats?

Conditions have improved in all three areas. Fish are no longer laced with the extreme concentrations of mercury and DDT found in the 1960s. Most public drinking water supplies are tested and treated at least for bacterial contaminants, and some also for toxics. And bacteria levels have declined in many of the nation's most polluted waters. But most people still cannot drink from their taps, eat the catch from their local fishing hole, or take their family to their favorite beach or lake with full assurance that they will not become ill or face long-term health threats.

Swimming Hazards and Beach Closures

Among the first questions the average person would ask about whether the Clean Water Act has succeeded are whether it is safe to go to the beach and whether swimming waters have improved over the past twenty years. Unfortunately, information on the safety of our swimming waters is scarce. But available data suggest that a disturbing number of waters around the country remain unsafe for swimming.

Swimmers can become ill from several microorganisms (see Table 2.4). Viruses are believed to be the major cause of diseases such as gastroenteritis and hepatitis. Complete data on illnesses from swimming in polluted waters are difficult to find, because so many of these illnesses go unreported. One study, however, estimated that as many as fifty-eight thousand illnesses a year among Hong Kong residents are caused by swimming at a single beach.[54] Closer to home, a study conducted for EPA from 1972 to 1979 in New York, Lake Pontchartrain, Louisiana, and Boston, Massachusetts, found "[a] direct, linear relationship between swimming-associated gastrointestinal illness and the quality of the bathing water. . . . Moreover, the ratio of the swimmer to non-swimmer symptom rates indicated that swimming in even marginally polluted marine bathing water is a significant route of transmission for the observed gastroenteritis."[55]

A more recent study, of lifeguards in Los Angeles County, identified other types of health problems from swimming in polluted waters.[56] Workers' compensation claimants were matched to healthy lifeguards who had the same job classification. A reported 112 acute compensable illnesses occurred among the lifeguards during the seven-year period (1980–86). Most of these illnesses affected lifeguards who worked at the southernmost beaches, which had high levels of bacteria from storm drains and other sources.

The logical way to evaluate whether more waters are safe for swimming now than in 1972 would be to analyze the numbers of beaches that have been closed each year. Incredibly, however, this information does not exist, at least not in one place. To the extent that historical beach closure information exists at all, it resides in hundreds of state and local offices around the country. Until NRDC began to collect beach closings data in 1989, there appears to have been no comprehensive effort to evaluate this issue at all. The Environmental Protection Agency's *1988 National Water Quality Inventory* was the first attempt to report on beach closures, and it provided only the most cursory information.[57]

The Environmental Protection Agency's *National Assessments* do report how many waters the states believe are "swimmable." In the *1990 Assessment,* for

TABLE 2.4

Pathogens and Swimming-Associated Illnesses

PATHOGENIC AGENT	DISEASE
Bacteria	
Vibrio cholerae	Cholera
Salmonella typhi	Typhoid fever
Other *Salmonella* species	Various enteric fevers (often called paratyphoid), gastroenteritis, septicemia (generalized infections in which organisms multiply in the blood stream)
Shigella dysenteriae and other species	Bacterial dysentery
Viruses	
Rotavirus	Gastroenteritis
Norwalkvirus	Gastroenteritis
Poliovirus	Poliomyelitis
Coxsackievirus (some strains)	Various, including severe respiratory disease, fevers, rashes, paralysis, aseptic meningitism myocarditis
Echovirus	Various, similar to coxsackievirus (evidence not definite except in experimental animals)
Adenovirus	Respiratory and gastrointestinal infections
Hepatitis	Infectious hepatitis (liver malfunction), also may affect kidneys and spleen
Protozoa	
Cryptosporidium	Gastroenteritis
Giardia lamblia	Diarrhea (intestinal parasite)
Entamoeba histolytica	Amoebic dysentery, infections of other organisms
Isospora belli and *Isospora hominus*	Intestinal parasites, gastrointestinal infection
Balantidium coli	Dysentery, intestinal ulcers

Source: NRDC, *Testing the Waters* (1992), Table 2.

example, EPA reported that the Clean Water Act's swimmable goal was met in about three-quarters of our rivers and estuaries, more than 82 percent of our lakes, and almost 90 percent of our ocean waters. Even these optimistic numbers lead us to conclude that, almost a decade after the 1983 goal for swimmable waters, a large number of water bodies (one out of ten ocean miles and one of five lake acres) are not safe for swimming. But a closer analysis indicates that many more waters are not safe for swimming.

First, there are wide variations in state water quality standards and monitoring programs (discussed in more detail in chapter 3) and in the criteria states use to determine whether a use has been met. More fundamentally, as discussed previously, even where EPA's recommended criteria are used, bacteria and other standards can be violated a significant number of times for waters deemed fully swimmable. For example, a water body can violate bacteria criteria in one out of every ten samples and still receive the highest (full support) rating. A person who swims on one of those days is simply a statistical casualty. Worse, since there are no distribution requirements in EPA's standard, a beach could violate criteria every day during a family's vacation week, yet still meet the one out of ten criterion over the entire year and be deemed swimmable.

More informative assessments of beach closures from 1989–92 are available in NRDC's *Testing the Waters: A National Perspective on Beach Closings,* issued in July 1992, and *Testing the Waters: The Price of Coastal Pollution and Beach Closings* (expected to be issued in July, 1993) (see Table 2.5). The major findings of this report indicate that we are indeed a long way from achieving the swimmable waters goals of the act:

● In 1991, U.S. ocean and bay beaches were closed or advisories were issued against swimming on more than two thousand occasions in coastal states that monitor beach water quality. There were over 2,600 closures or advisories in 1992 (and probably more, as many key areas did not provide closings data to NRDC for 1992). High bacteria levels were responsible for the overwhelming majority of cases. More than 7,700 closures or advisories have occurred since 1988.

● Most of the 1991 closures and advisories (715) were in the densely populated coastal areas of New York, New Jersey, and Connecticut; 588 others were in southern California from San Diego to Los Angeles. Similarly, in 1992, over 1,100 closures and advisories were reported in New York, New Jersey and Connecticut, and over 600 were in California.

● States have highly inconsistent water quality standards for sewage contamination and inadequate (and largely nonmandatory) beach closing standards and criteria.

- Lack of federal leadership has resulted in the complete absence of monitoring in some states and in substantial variations in testing methods and closure standards. Only four states use EPA's recommended testing method.

- Levels of protection vary. Among twenty-two states surveyed, eleven different bacteria standards were used for primary water contact; in seven states, standards even vary within the state.

It is instructive to compare this beach closure information with the numbers of waters reported by states to violate the swimmable waters goal of the Clean Water Act in the *1990 Water Quality Inventory*. This analysis confirms that the section 305(b) reporting system seriously understates the safety of the nation's coastal waters for swimming. In some cases, state reporting appears to be based on little or no actual monitoring data; in others, the reports seem directly inconsistent with the numbers of beach closures found by NRDC. For example:[65]

- Alabama conducts no regular monitoring of marine beaches, yet it reports that no estuaries and no coastal waters fail to meet the swimmable goal of the Clean Water Act. Since 28 percent of Alabama's shellfish beds were closed and a total of 85 percent were closed or harvest limited in 1990 due to bacteriological pollution, it is incredible that none of these waters is unsafe for swimming as well.

- Delaware reported that none of its ocean waters failed to meet the swimmable goal of the act in 1988–89; yet NRDC's survey identified sixty-two beach closures in 1989.

- Georgia conducts no regular monitoring of marine beaches and actually has no bacteria standards to protect swimmers in coastal waters. Accordingly, Georgia reports no information on coastal waters that fail to meet the swimmable goal of the act. In 1990, however, 72 percent of Georgia's shellfish waters were restricted or closed due to bacteria, indicating probable swimming hazards as well.

- Hawaii had at least nine beach closures in 1988 and at least twenty-three in 1989. Moreover, Hawaii only closes beaches due to sewage spills. For example, in 1990, Hawaii beaches violated the state bacteria standard for swimming 342 times, but only twenty-two closures or advisories were issued. Despite this ongoing pollution, Hawaii reported that no coastal or estuarine waters failed to meet the swimmable goal of the Clean Water Act in 1988 and 1989.

- Mississippi ended its beach-monitoring program in 1989 and reported that no coastal or estuarine waters fail to meet the swimmable goal of the act. Yet

TABLE 2.5

Ocean and Bay Beach Closures and Advisories in Twenty-two Coastal States, 1988–92				
STATE	1988	1989	1990[58]	1991	1992
Alabama*	—	—	—	—	—
California[59]	**	At least 64	At least 338	745 + 5 permanent, 1 extended	At least 609 + 1 permanent, 2 extended
Connecticut	**	At least 103	218	293 + 1 extended	At least 223
Delaware	1	62	11	11	5
Florida[60]	**	**	303	299	773 + 1 extended
Georgia*	—	—	—	—	—
Hawaii[61]	At least 9	At least 23	At least 22	106	29
Louisiana	***	—	—	1 permanent	1 permanent
Maine	**	1	30 + 1 extended	47 + 3 permanent	At least 3 permanent
Maryland	0	0	0	24 + 3 permanent, 2 extended	At least 6 + 3 permanent, 2 extended
Massachusetts[62]	At least 75	At least 60	At least 59	At least 59	At least 60
Mississippi[63]*	—	—	—	—	—
New Hampshire****	—	—	—	1 extended	0

monitoring in 1989 found that bacteria standards were exceeded at eight out of eleven recreational beaches, with average summer bacteria levels more than 50 percent above the state standard.

● New York has the highest number of beach closures and advisories in the country, with 273 in 1988 and 473 in 1989. Yet in its 305(b) report, New York claimed that all of its assessed ocean waters met the swimmable goal of the act, as did the vast majority of estuarine waters (1,487 out of 1,578 square miles).

(TABLE 2.5 CONTINUED)

STATE	1988	1989	1990[58]	1991	1992
New Jersey	126	266	228	108	112
New York	273 + 1 permanent	473 + 5 permanent	383 + 3 permanent	314 + 3 permanent, 2 extended[64]	799
North Carolina*	—	—	—	—	—
Oregon*	—	—	—	—	—
Rhode Island	0	0	0	0	0
South Carolina*	—	—	—	—	2
Texas*	—	—	—	0	1 medical advisory
Virginia	**	**	**	2	0
Washington*	—	—	—	—	—
TOTAL	At least 484 + 3 permanent	At least 1,052 + 5 permanent	At least 1,592 + 4 permanent, 1 extended	At least 2,008 + 14 permanent, 7 extended	At least 2,619 + 8 permanent, 6 extended

Source: NRDC, *Testing the Waters* (1992), Table 1; *Testing the Waters* (1993), Table 1.
Note: A beach closure/advisory is a single beach for which a closure/advisory has been issued for a single day.
Permanent closures were for at least the entire summer, while extended closures were for more than six weeks.
* This state does not monitor marine beaches regularly.
** NRDC did not gather data for this year.
*** This state does not monitor marine beaches regularly. It had 1 permanent closing.
**** This state does not monitor marine beaches regularly. It had 1 extended closing.

Because of inconsistencies in monitoring and closure practices among states and over time, it is impossible to make comparisons among states or to assess trends based on the closure data.

● Texas has no regular monitoring of marine beaches, yet it reported that all of its estuaries and all but 1 mile of its ocean waters meet the swimmable goal of the Clean Water Act. In 1990, more than half of the shellfish waters in Texas were restricted due to the same pollutants that can pose health risks to swimmers.

Based on these large data gaps and inconsistencies, it is clear that we do not have adequate information on how many of our coastal waters meet the swimmable goal of the Clean Water Act. It is equally clear that we cannot simply rely on biennial state water quality reports to present complete and accurate information on this problem.

The Environmental Protection Agency acknowledges the serious inconsistencies among states in standards to protect swimmers. In an effort to correct this problem, the agency is considering convening a policy dialogue among affected interest groups (recreationists, environmental groups, state, local, and the federal government, and so on) to discuss ways to make beach-closing standards and procedures more consistent around the country. In the meantime, however, swimmers will continue to remain uncertain about the relative safety of their favorite freshwater and coastal beaches.

Environmental Justice and Beach Pollution

A particularly disturbing aspect of beach pollution is that the most polluted waters tend to be in or near densely populated cities, particularly older cities with combined sewer overflows (CSOs) and other discharges of raw or partially treated sewage. While some people can afford to avoid this problem by traveling to cleaner but expensive resort areas, others have neither the time nor the money to do so. As a result, low-income city residents, often disproportionately African Americans, Hispanic Americans, and other people of color, are forced to choose between swimming in polluted waters or not swimming at all.

City and state officials, in turn, must choose between protecting public health and closing the only beaches realistically available to some city residents. Eric Mood, a professor of public health at Yale University, foresaw the beach equity issue as early as 1970, when officials in Connecticut reportedly kept polluted beaches open to avoid riots in urban areas: "As a professional health official [he] knows he should close the beach for swimming. But as a humanitarian he knows that the socially deprived residents of his city have no other convenient outdoor bathing area to which they may seek relief from the oppressing summer heat of the urban slums in which they live."[66]

More recent and more careful studies have confirmed that differences exist in outdoor recreation activities, whether due to economic or other factors. As a result, eliminating pollution of urban recreational waters is particularly important to people of color. Studies of outdoor recreation patterns in Illinois, for example,[67] showed that whites are more likely than other groups to recreate at private clubs that require membership, such as swim clubs or country clubs. Whites are also more likely to take overnight trips in or out of state. "There is less overnight travel by Blacks, Hispanics, and Asians; and their concentration in urban areas (particularly Blacks) makes urban and near-urban resources especially critical to these groups."[68]

These and other studies led EPA officials to conclude that CSO discharges disproportionately affect certain populations and that pollution of city waters may severely limit access by low-income residents, including African Americans, Hispanic Americans, and other groups, to water-based recreation. Some information suggests that these groups decide to swim in these polluted waters anyway, either because they do not know the health risks or because they lack alternatives.[69]

Pollution of Drinking Water

Another basic question that an average citizen is likely to ask about the success of water pollution control efforts is whether the drinking water that comes out of their taps is safer to drink than it was in 1972. Information is inadequate for a full assessment of the safety of our drinking water supplies. But the data that are available suggest that millions of people in the United States face health risks due to contamination of surface water and ground water.

The question "Is Your Water Safe?" was splashed across the front cover of *U.S. News and World Report* on July 29, 1991. The article began:

[N]early two decades after Congress passed the Safe Drinking Water Act, the water flowing from the nation's taps is anything but pristine. One in 6 people drink water with excessive amounts of lead, a heavy metal that impairs children's IQ and attention span. In the early summer, half the rivers and streams in America's corn belt are laced with unhealthy levels of pesticides. Microbes in tap water may be responsible for 1 in 3 cases of gastrointestinal illness. Ironically, even the chlorine widely used to disinfect water produces carcinogenic traces when combined with other common substances in water.

The article mentions the Safe Drinking Water Act because that law regulates the quality of water as it leaves your tap, while the Clean Water Act is designed to eliminate water pollution in the rivers and lakes from which half of the country (by population) gets its drinking water. But while the general public does not care about this fine legal distinction, progress under the Clean Water Act is critical to the average citizen's drinking water for two reasons: First, ultimately, cleaner water supplies will produce cleaner water to drink. Second, the public faces increasing costs for treating drinking water to eliminate contaminants that should not be there in the first place. (We will discuss the high costs of treating polluted water rather than preventing the pollution to begin with in chapter 3.)

Relying on treatment of polluted water supplies to protect public health instead of eliminating the pollution in the first place is dangerous, as evidenced by major disease outbreaks caused by problems in public water treatment systems. In 1987, thirteen thousand people in Carrollton, Georgia, contracted Cryptosporidiosis—a gastrointestinal illness, caused by the protozoan Cryptosporidium, that can be life threatening to people with immune deficiency, such as AIDS sufferers and chemotherapy patients—from a filtered public water supply that reportedly met both federal and state drinking water standards. Officials postulated that the Cryptosporidia were from livestock runoff and had passed through apparently faulty and improperly maintained equipment.[70] This same protozoan attained national notoriety in 1993 when at least 183,000 people in Milwaukee, Wisconsin became ill from drinking unfiltered public water; again, health officials cited livestock runoff as one likely source of the pathogens.[71] In 1985, about thirty-eight hundred people in Pittsfield, Massachusetts, contracted Giardiasis from local reservoir water. The pathogens passed through the treatment system, reportedly due to a malfunction in chloridation (disinfection) equipment.[72] A recent survey of the raw water supply for 66 water treatment plants in fourteen states and one Canadian province found either Cryptosporidia or Giardia in 97 percent of the samples, and a high correlation between the presence of these organisms and industrial and sewage discharges and other indications of poor water quality.[73]

The only reliable, reasonably comprehensive information on trends in drinking water quality is available from the Centers for Disease Control and Prevention, which keep statistics on waterborne disease outbreaks.[74] These reports identified 554 disease outbreaks related to public water supplies from 1972 to 1990, affecting almost 136,000 people. Between 1971 and 1985, more outbreaks of disease were reported than in any previous fifteen-year period since 1920.[75] Even these numbers are conservative; CDC acknowledges that the number of waterborne disease outbreaks reported to CDC and EPA represents only a fraction of the total number that occur. EPA reports that some researchers believe there are twenty-five times more actual illnesses from contaminated drinking water than are reported.[76] If correct, this would translate to illnesses affecting more than 3.4 million people for the period from 1972 to 1990.

The number of outbreaks per year generally increased between the early 1970s and the early 1980s and then dropped again between 1984 and 1990. Total illnesses differ widely from year to year, reflecting high variations in the numbers of people affected per outbreak. This involves not just the severity of the contamination incident but the size of population served by the drinking water system. Notably, the highest number of total illnesses in a single year (until the 1993 Milwaukee outbreak) was in 1987, indicating that serious problems remain due to contamination of drinking water by pathogens (see Table 2.6).

Aside from these reports about contamination by pathogens, little historical information is available to evaluate how much the quality of drinking water has changed over the past twenty years. Two other types of data are available, however, to begin to answer the question of how safe drinking water is today.

The first approach to use in answering this question is to evaluate available data on what contaminants are present in our drinking water supply, and at what levels.[77] Some of the information is alarming. For example, a 1984 survey conducted by the National Cancer Institute identified a list of 1,565 organic chemicals in drinking water, 117 of which were known to cause health effects at that time.[78] A 1990 report by the GAO listed sixty-six common contaminants in drinking water, including both chemicals and disease-causing organisms, along with their associated health effects. The effects ranged from intestinal disorders, to various forms of cancer, to developmental effects, to a wide range of chronic ailments involving the central nervous system, heart, liver, kidney, and respiratory system.[79]

No easily accessible source of information quantifies current levels of contaminants in drinking water on a national basis, much less evaluates trends over the past twenty years. Public drinking water supplies are required to monitor for chemical contaminants and pathogens,[80] and these data are available on the Federal Reporting Data System (FERDS). However, the data have not been analyzed or summarized in a usable form for purposes of a national assessment.

The most useful nationwide perspective comes from an analysis of drinking water standards violations around the country, as reported in EPA's most recent national summary.[81] The maximum contaminant level (MCL) compliance rate for community water systems (CWS) remained between 70 percent and 73 percent from 1986 to 1991. This figure represents the number of MCL violations reported to EPA. In other words, between 27 percent and 30 percent of community drinking water systems reported violations of health-based standards during these years. For some perspective on the causes of these problems, in 1991 there were 4,887 violations for microbiological contaminants, 285 turbidity violations, 165 violations involving inorganic chemicals, 89 violations for organic chemicals, and 97 radiological contaminant violations. These data may be misleading, however, because many states did not report certain types of violations (such as certain chemicals or radiological contaminants) to EPA.

TABLE 2.6

Waterborne Disease Outbreaks—United States, 1972–90

YEAR	TOTAL OUTBREAK	TOTAL CASES
1972	30	1,650
1973	25	1,762
1974	25	8,356
1975	24	10,879
1976	35	5,068
1977	34	3,860
1978	32	11,435
1979	45	9,841
1980	53	20,045
1981	36	4,537
1982	44	3,588
1983	43	21,036
1984	27	1,800
1985	22	1,946
1986	22	1,569
1987	15	22,149
1988*	16	2,172
1989-90	26	4,288
TOTAL	554	135,981

Source: U.S. Department of Health and Human Services, *Waterborne Disease Outbreaks 1986–88; Foodborne Disease Outbreaks, 5–Year Summary, 1983–87*; B.L. Herwaldt, G.F. Craun, S.L. Stokes and D.D. Juranek, "Outbreaks of Waterborne Disease in the United States: 1989–90," *Journal of the American Waterworks Association,* April 1992, 129–35.
* Includes three additional outbreaks affecting 44 people not reported in the 1986–88 summary, but identified in the 1989–90 report.

According to EPA's data, both the total number of MCL violations by CWSs and the number of such systems in violation apparently declined somewhat from 1986 to 1988, while the number of CWSs in EPA's inventory increased steadily. Between 1988 and 1991, however, both the number of violations and the number of systems in violation rose. EPA believes this was due to the implementation of new regulations and more complete reporting. An NRDC report scheduled for release later this year, however, will conclude that the number of MCL and monitoring and reporting violations by all public water systems has increased since 1986, according to a more detailed analysis of available data. The number of significant noncompliers (CWSs that have more serious, frequent, or persistent violations) declined by more than half between 1986 and 1990, but analysis of this apparent trend is seriously complicated by EPA's redefinition of what constitutes a "significant noncomplier."[82]

EPA's violation figures, generally, count only violations in community water systems, which serve year-round residents. Noncommunity water systems serve primarily nonresidential areas or facilities without year-round residents. Of the approximately 201,000 public water systems in the United States, EPA classifies roughly 59,000 as community systems and about 142,000 as noncommunity systems. Studies conducted by the National Wildlife Federation (NWF) in 1988 and 1989,[83] using searches of EPA's computer records, showed much higher numbers of drinking water violations, affecting millions of people, when both community and noncommunity systems were counted (see Table 2.7).

Even assuming that EPA is correct and that drinking water violations are declining somewhat, it is difficult to determine whether this results from bet-

TABLE 2.7

Public Drinking Water System Violations—1987–88				
	COMMUNITY		NONCOMMUNITY	
	1987	1988	1987	1988
Systems in violation	16,250	15,616	20,513	18,574
Total violations	49,641	48,507	51,947	48,986
Population affected	37.4	38.4	3.6	3.1
(in millions)				

Sources: N.L. Dean, *Danger on Tap: The Government's Failure to Enforce the Federal Safe Drinking Water Act* (National Wildlife Federation, 1988); National Wildlife Federation, *Danger on Tap: FY 1988 Update* (1989).

ter treatment or from cleaner water supplies. In any event, the large number of continuing violations indicates that many U.S. waters are no more drinkable than they are swimmable, even after expensive treatment.

Fish and Shellfish Contamination

Available information shows that levels of contaminants in much of our seafood are declining as we reduce the amount of toxic pollutants we dump into our waters. But many consumers still face serious health risks from some sources of fish and shellfish, particularly in heavily polluted urban and industrial waters.

The general public also has a strong interest in the safety of our seafood supply. Because seafood is healthful—one of the best, low-fat sources of protein—Americans are eating fish in record amounts; average annual seafood consumption jumped from about 12.5 pounds per person in 1972 to 15.5 pounds in 1990. The people of this country, especially in coastal areas, also have a long and rich history of enjoying seafood for cultural and aesthetic reasons.

In February 1992, *Consumer Reports* asked the U.S. public the same question about fish and shellfish that *U.S. News and World Report* had asked about drinking water a year earlier: "Is Our Fish Safe to Eat?" The Consumers Union survey reported that almost 40 percent of fish were of fair or poor quality, and nearly 30 percent of those were of poor quality. Nearly half were contaminated by bacteria from human or animal feces, although the article reported that this was most likely from poor sanitation practices during fish handling, that is, not from polluted waters. Moreover, the article reported that thoroughly cooked fish, even if contaminated, won't make you sick.[84]

The article also found high percentages of fish and shellfish with chemical contaminants, which do occur as a result of water pollution:

- 43% of salmon samples contained PCBs, which may cause cancer and pose reproductive hazards.
- 90% of swordfish samples had mercury, which may harm the nervous system; 25% contained PCBs.
- Some catfish had residues of the pesticides DDT, DDE, and DDD, which can affect mammalian reproduction.
- Some clams were high in lead, which can impair behavioral development in young children.
- 50% of lake whitefish had PCBs; some had trace pesticides.
- Some fish, on the other hand, such as flounder and sole, were virtually free of pollutants.[85]

Are these types of seafood contamination real, or is this media hype? Is the situation getting better or worse because of Clean Water Act pollution controls? The answers lie somewhere in between. While the majority of seafood that reaches U.S. tables is relatively safe, there is considerable cause for concern about seafood caught from polluted waters. In its guidance manual for evaluating health risks from seafood, EPA warns:

> Contamination of aquatic resources by toxic chemicals is a well recognized problem in many parts of the U.S. High concentrations of potentially toxic chemicals have been found in sediments and in aquatic organisms from Puget Sound, the Southern California Bight, northeast Atlantic coastal waters, the Hudson River, the Great Lakes, and elsewhere. Heavy consumption of contaminated fisheries products by humans may pose a substantial health risk. This concern has prompted recent studies of catch and consumption patterns for recreational fisheries and associated health risks.[86]

Evidence from around the country indicates that levels of many dangerous contaminants are declining in parallel with reduced releases of toxic chemicals, but some remain above levels at which serious health effects can occur. In some areas, especially those with pollution caused by raw or partially treated sewage, the situation appears to be getting worse. And unfortunately, for both the public and the seafood industry, an increasing number of waters are being closed to commercial and recreational fishing.

Fish Consumption Advisories and Bans

In 1991, the National Institute of Medicine released a comprehensive review of available information on seafood contamination.[87] The study concluded that, while most seafood consumed in the United States is safe and a valuable part of a healthy diet, serious risks remain, particularly from consumption of raw shellfish and from both artificial and natural toxins in fish and shellfish. Subsistence and recreational fishers were found to be at particularly high risk due to high rates of consumption and high contaminant levels in some freshwater fish. Following are more detailed discussions of the principal data sources evaluated in this study and of others available after its release.

The presence of toxic pollutants in seafood is both a sensitive measure of past and present pollution and a cause for serious human health concern due to the concentration of these chemicals[88] in the tissue of aquatic life through *bioaccumulation*. Bioaccumulation occurs in two ways. Fish accumulate toxics from the water column directly through the skin and gills, a process called *bioconcentration*. Further uptake occurs through the food chain; this process is known as *biomagnification.* Any judgment of human exposure to toxic pollutants based on pollutant levels in the water itself (as opposed to more direct measures of toxics in seafood) must properly take these factors into account.

Information on the degree of seafood contamination is available from a number of sources. One way to evaluate ongoing levels is to consider the number of fishing advisories or bans around the country. In the *1990 National Water Quality Inventory,* thirty-one states reported concentrations of toxic contaminants in fish at levels exceeding those known or suspected to pose human health risks. Forty-five states reported almost one thousand fishing advisories and fifty complete fishing bans in 1988–89, due to pollutants such as PCBs, pesticides, dioxin, mercury and other metals, and other organic chemicals. These warnings affected more than 7,000 river miles, almost 2.5 million lake acres, more than 800 square miles of estuaries, and almost 5,000 miles of shoreline in the Great Lakes.[89]

Table 2.8 shows that more states are collecting information on fish contamination than were a few years ago and as a result are issuing more bans and advisories affecting more waters. But EPA readily acknowledges that these reports are still incomplete and that state standards for deciding whether such advisories are appropriate vary widely.[90] We discuss the lack of adequate monitoring for fish contamination in more detail in chapter 4.

The 305(b) report itself appears incomplete in its identification of fish advisories known to EPA. A review of EPA's computer database of state fish advisories conducted by the Environmental Defense Fund (EDF) revealed nearly four thousand advisories for one or more toxic pollutants,[91] four times the

TABLE 2.8

	State-Reported Fishing Restrictions						
	STATES REPORTING	BANS	ADVISORIES	RIVER (MI)	LAKE (ACRES)	ESTUARIES (SQ MI)	GREAT LAKES SHORELINE (MI)
1988–89	45	50	998	7,226	2.5 mil	836	4,808
1986–87	47	135	586	5,487	2.8 mil	334	4,098
1984–85	25	108	286	—	—	—	—
1982–83	24	42	88	—	—	—	—

Source: EPA, National Water Quality Inventories (1990, 1988, 1986, 1984).

number reported in the *National Water Quality Inventory*. The pollutants most frequently responsible for these warnings include PCBs, dioxins, pesticides, and metals such as mercury, lead, and selenium.

Even this higher number of state fish consumption advisories, however, most likely understates health risks to the public. Most states decide whether a fishing advisory is appropriate based on "action levels" for specific pollutants established by the FDA, which are levels at which FDA takes enforcement action to remove commercial seafood from the market. But action levels are available only for a handful of pollutants. Moreover, these levels do not address the full range of human health effects, are based on outdated science, were set to protect the general public from commercial seafood in interstate commerce, and assume that each "average" consumer eats a broad mix of species from diverse sources. They are not designed to protect recreational or subsistence fishers, who eat much larger amounts of fish from a single or a small number of contaminated waters. Nor do they account for the cumulative effects of multiple pollutants. These and other factors have led the FDA itself, as well as EPA, the National Institute of Medicine, the National Oceanic and Atmospheric Administration (NOAA), and others, to criticize state use of FDA action levels to issue fish advisories for local waters.[92] Use of more appropriate assumptions about local consumption patterns would result in many more fishing bans and advisories around the country.

Because of these limitations, and because data and procedures for issuing fish advisories have changed over time, the number of advisories issued each year is not useful in determining whether seafood safety has improved since 1972. National databases are available, however, to evaluate general trends in seafood contamination. These include EPA's National Study of Chemical Residues in Fish, the U.S. Fish and Wildlife Service's (FWS's) National Contaminant Biomonitoring Program (NCBP) for inland waters, NOAA's National Status and Trends and Mussel Watch Programs for coastal waters, and the National Shellfish Register for shellfish beds. All of these sources have problems, but they paint a reasonable portrait of overall contaminant trends.

EPA's National Study of Chemical Residues in Fish

In late 1992, EPA released the results of a five-year effort to evaluate the presence of toxic chemicals that may be bioaccumulating in fish.[93] This effort began as the National Dioxin Study, a nationwide investigation of the most toxic form of dioxin. The presence of dioxin in fish at unexpected levels caused EPA to launch a broader evaluation of the presence of other potentially bioaccumulative chemicals in fish around the country. This effort, originally called the National Bioaccumulation Study, tested for the presence of sixty pollutants in 119 species of fish collected from 314 water bodies. Many test sites were chosen based on predicted impacts from point sources or sources of polluted runoff, but others were selected to reflect background conditions.

The results of the EPA study are sobering. Almost half of the chemical forms of dioxins and furans, and one-third of the other chemicals measured, were found at more than half of the sampling locations. Biphenyl, mercury, PCBs, and DDE were found at more than 90 percent of the test sites. And every pollutant in the study was found in at least one location.

Concentrations of pollutants varied widely among individual samples. Nevertheless, EPA calculated that the levels of pollutants measured in fish around the country posed significant risks of cancer and other health effects to average fish consumers and even higher risks to subsistence and recreational anglers who consume more fish from contaminated waters. (Chapter 4 includes a more detailed discussion of the way EPA calculates cancer and other health risks.)

The National Study of Chemical Residues in Fish has obvious limitations in terms of establishing national trends. Despite the national scope of the study, only a limited number of pollutants were tested at a limited number of sites. Moreover, the study was not designed to establish trends and includes no time sequence data. Nevertheless, the EPA report presents the most current and comprehensive evidence that a wide range of toxic chemicals contaminates freshwater fish around the country at levels that pose severe human health threats.

Moreover, the results of EPA's recent study are largely consistent with the FWS's National Contaminant Biomonitoring Program. While differences in sample locations and pollutants measured make direct comparisons impossible, the persistent presence of contaminants in harmful quantities indicates that we remain far from our goal of eliminating the release of toxic pollutants into the aquatic environment, and we continue to face health threats from pollutants released many years ago.

FWS's National Contaminant Biomonitoring Program

Two recent reports summarize NCBP findings, tracing contaminant levels of toxic metals and organic chemical pollutants respectively in inland waters from 1976 to 1984.[94] The data showed many victories but significant remaining battles. Levels of toxic metals in fish tissue declined for some metals (arsenic, cadmium, copper, zinc, and selenium) from the mid-1970s to the early 1980s and then leveled off. Mercury levels had already declined by more than 25 percent between the late 1960s and 1974 but showed no appreciable change thereafter. The pollutant that showed the most consistent decline was lead, in conjunction with the phaseout of lead as a gasoline additive. On average, no pollutants showed increasing concentrations.

Despite this general decline, however, detectable levels of toxic metals remained in most fish sampled, often at levels of health concern. While there is considerable difference of opinion about safe levels of toxic pollutants in fish, EPA's seafood contamination guidance manual presents a range of criteria used by various countries.[95] Average measured concentrations in the NCBP exceeded at least the low end of this range for all of the tested metals except copper, lead, and zinc; maximum detected levels were well into this range for all of the metals. For all of the pollutants except cadmium and copper, contaminant levels were increasing in one or more monitoring stations; levels of zinc increased at one-third of all stations.

Similar trends were apparent for pesticides and other chlorinated organic chemicals, such as PCBs. Concentrations of some of these pollutants appeared to be declining, while others remained fairly constant or declined initially and then leveled off. While these data show some improvement, substances such as PCBs, DDT, and other pesticides continued to be found in fish tissue long after the use of these chemicals was banned. Average contaminant concentrations remained at levels of concern for the pesticides endrin, chlordane, and toxaphene (again measured against the low end of EPA's reported range of criteria used by various countries), and the highest detected levels remained high for most of the pollutants.

While the NCBP was an extremely useful source of national data, it also had serious limitations in terms of measuring toxics in fish. The monitoring stations were well distributed around the country, but the overall system was small relative to the large number of potentially contaminated waters. And fish were tested for only a small number of pollutants, many of which are pesticides or other chemicals (such as DDT and PCBs) that are no longer in use.

NOAA's National Status and Trends and Mussel Watch Programs

A similar monitoring network, maintained by NOAA, measures sediment and fish and shellfish contamination in the nation's estuaries and other coastal waters. Like the FWS network, this program is limited to a relatively small number of pollutants and does not include many industrial or agricultural chemicals currently in use. Moreover, because it is designed primarily to address ambient conditions in broad areas,[96] the monitoring locations selected were not the most contaminated areas and in fact were chosen to avoid heavily polluted sites. Officials of NOAA note that site-specific studies often detect much higher contamination levels than the broader NOAA program does, due to past or ongoing pollution of urban and industrial areas.[97]

Like the FWS program for inland waters, NOAA's coastal monitoring indicates much improvement but significant, and in some cases deteriorating, problems. On a national scale, the Status and Trends and Mussel Watch Programs[98] show that contaminant levels are decreasing for most of the pollutants measured. Levels of most toxic metals were either relatively level or decreasing. However, the only pollutant for which all sites showed decreasing levels was PCBs (which are no longer manufactured in the United States); some sites improved and others worsened for other pollutants. In the cases of lead, mercury, and selenium, in fact, more sites deteriorated than improved, and copper showed an increasing trend overall. Notably, copper is the only metal tested for which overall national use has increased during the period of this program.

While the monitoring locations were not selected to detect toxic "hotspots," the most polluted areas—and the areas most likely to show increasing contamination—were in heavily populated urban and industrial areas. These include sites near Boston; New York; San Diego; Los Angeles; and Seattle. Water bodies with particularly high levels of contamination include Long Island Sound; the Hudson/Raritan Bay Estuary; portions of the Chesapeake Bay; and Puget Sound.

Thus, as in inland waters, chemical contamination of fish appears to be generally decreasing in coastal waters. But progress is still slow relative to the Clean Water Act goal of fishable waters by 1983, and problems are actually getting worse in some highly polluted areas and for some pollutants.

The National Shellfish Register

The overall level of contamination of coastal waters by sewage and other sources of pathogens, unlike chemical contaminants, appears to be getting worse (although this probably reflects better monitoring and reporting as well as ongoing pollution). The National Shellfish Register and CDC and FDA data on foodborne disease outbreaks both show this trend.

The National Shellfish Register,[99] for example, shows that the amount of estuarine waters in which shellfishing was banned increased by 6 percent from 1985 to 1990 (see Figures 2.2A–2.2F). By 1990, in fact, less than two-thirds of our shellfish waters were unconditionally approved for shellfish harvest. On a regional basis, the situation was even worse. Between 1985 and 1990, the percentage of waters in which shellfishing was banned jumped by 10 in the Gulf of Mexico; it nearly tripled (from 10 to 29) in the North Atlantic. (While shellfish waters were degrading on the East and Gulf coasts, however, they appear to have been improving along the Pacific.)

FIGURE 2.2A

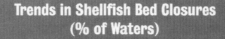

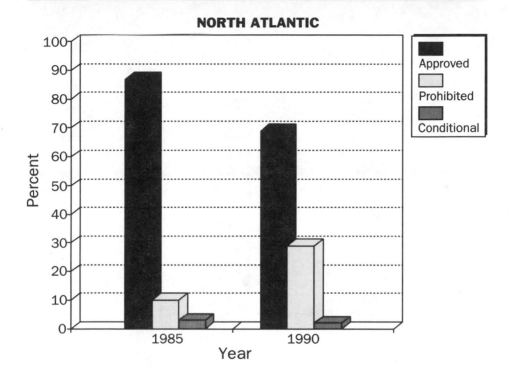

These data are hard to reconcile with reports in the *1990 National Water Quality Assessment* that more than two-thirds of all estuarine waters fully meet the fishable goal of the Clean Water Act. If state reporting of waters not meeting the fishable goal of the Clean Water Act were complete, the percentage of estuarine waters (where most shellfish beds are located) reported by the states as not meeting the fishable goal should at least roughly match the percentage of closed shellfish waters; and the percentage of estuarine waters not meeting or partially meeting the fishable goal (combined) should at least roughly match the total percentage of shellfish waters that are either closed or restricted. Although there may not be a 100 percent correlation between these two

FIGURE 2.2B

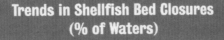

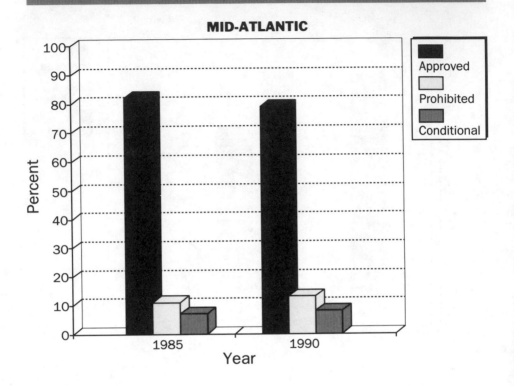

categories of waters, they should be reasonably close. Any differences in classification cannot explain the severe differences in the affected waters shown in Figure 2.3. In fact, these numbers are seriously inconsistent:

Some regions stand out in particular. For example, while more than one-fifth of the shellfish waters in the Southeast and more than one-third of the shellfish waters in the Gulf of Mexico are closed to shellfishing, Alabama, Mississippi, and Louisiana report no estuarine waters that fail to meet the fishable goal of the Clean Water Act, and South Carolina and Georgia report only 1 and 2 square miles of impaired estuaries, respectively. Clearly, judged by the percentage of estuarine waters closed to shellfishing, most states seriously

FIGURE 2.2C

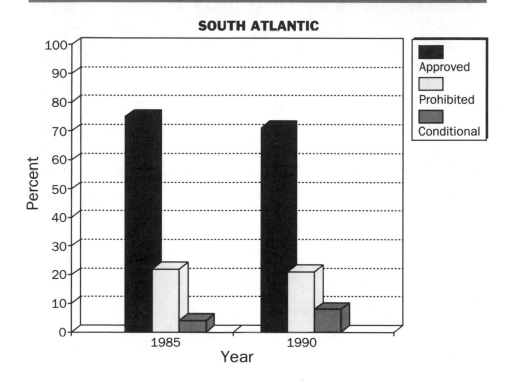

Trends in Shellfish Bed Closures
(% of Waters)

SOUTH ATLANTIC

underreport the amount of waters that fail to meet the fishable goal of the Clean Water Act.

This high level of contamination of coastal waters by sewage and other sources of bacteria and other biological contaminants is consistent with high reported levels of seafood-associated disease outbreaks, although "it is likely that only a small fraction of seafood-associated disease is reported and that the two available databases therefore reflect only a small fraction of the actual number of seafood-associated illnesses that occur."[100] An FDA database (NETSU) includes 5,342 cases of seafood-borne illness from 1978 to 1987, while CDC reported 3,271 shellfish-related cases and 203 other seafood-related

FIGURE 2.2D

Trends in Shellfish Bed Closures
(% of Waters)

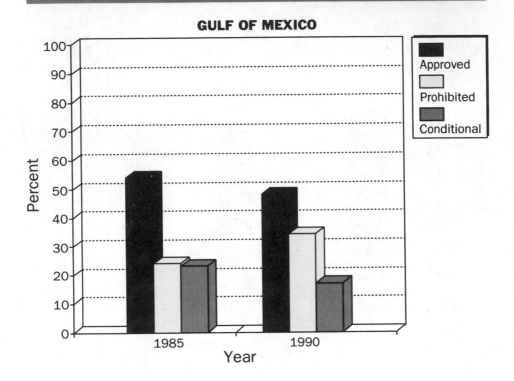

cases during same period.[101] Based on available data, consumption of raw shellfish is the most frequent source of illness.

Sediment Contamination

The NOAA Status and Trends and Mussel Watch Programs measure both sediment and fish contaminant levels at each monitoring site for good reason: There is a strong correlation between contamination of the sediment on the floor of coastal and other water bodies and contamination of the fish and shellfish that reach our tables.[102] Toxic pollutants in sediment contaminate

FIGURE 2.2E

Trends in Shellfish Bed Closures
(% of Waters)

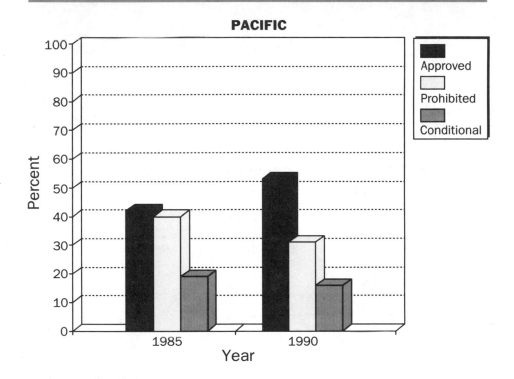

small aquatic organisms that live or feed in the sediment. These small animals are consumed by bottom-feeding fish, which in turn are eaten by larger fish. Bottom-dwelling shellfish can accumulate even higher levels of toxics because they filter-feed smaller organisms right out of the water. In fact, because levels of toxic pollutants bioaccumulate or biomagnify in higher levels of the food chain, sediment contamination levels can actually understate concentrations of the same pollutants in fish and shellfish.

In 1992, the Coast Alliance prepared a comprehensive survey of available information on sediment contamination.[103] Based on studies and compilations by EPA, NOAA, and the National Research Council (NRC), this review noted

FIGURE 2.2F

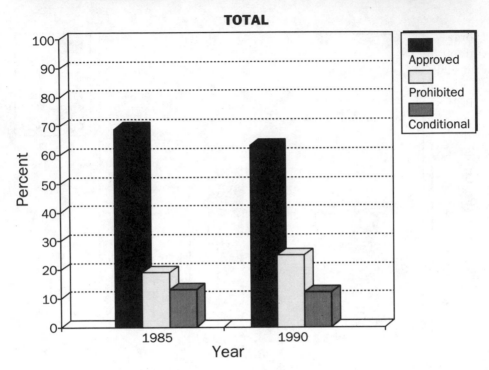

Source: NOAA, *The 1990 National Shellfish Register of Classified Estuarine Waters* (1991), Table 3.

hundreds of problem sites throughout the Atlantic, Gulf, and Pacific coasts, and the Great Lakes. The International Joint Commission (IJC) has identified serious contaminated sediment problems in 42 of 43 Areas of Concern in the Great Lakes and their tributaries. The nation's waters have become so polluted that, according to the EPA, only the most remote waterbodies can be expected to have pristine sediments.[104]

The Coast Alliance report identified and mapped fifty areas along the Atlantic, Pacific, and Gulf coasts that NOAA determined had sediment contamination at levels that threaten aquatic life, and an additional forty-three sediment hotspots in the Great Lakes.

The existence of contaminated sediment hotspots around the country in and of itself tells little about trends in sediment contamination, much less about the release of toxic pollutants into the marine environment. The only consistent national databases useful in evaluating sediment contaminant trends are the NOAA programs discussed previously. As with contamination of seafood, these programs show that concentrations of many toxic metals and organic chemicals are declining on a national scale. But sediment contamination is increasing for some pollutants, such as copper, and for other pollutants at some sites, especially in urbanized and industrialized areas.

Moreover, the persistent contamination of sediments by pollutants such as PCBs and certain pesticides, the U.S. manufacture of which was banned years

FIGURE 2.3

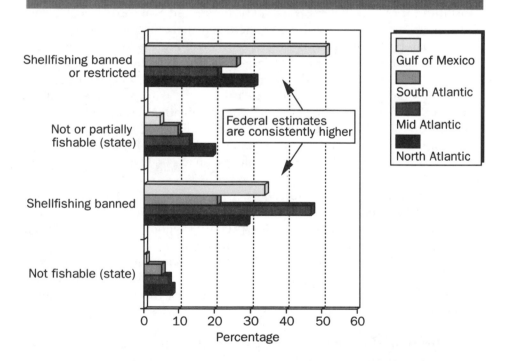

Comparison of Closed or Restricted Shellfish Waters to Estuaries Reported as "Unfishable" (% of Waters)

Sources: NOAA, *The 1990 National Shellfish Register of Classified Estuarine Waters* (1991), Table 3; EPA, *National Water Quality Inventory* (1990), 54.

* The following states did not report whether they met their fishable goal in estuaries: DE, NH, NJ, and NC.

ago, indicates that reductions in the contamination of sediment (as well as fish and shellfish) lag many years behind efforts to reduce or eliminate the release of toxic chemicals into our coastal and inland waters. This is true both because many of the toxic chemicals released into our waters are persistent—that is, do not degrade due to biological or other factors—and because sediments act as "sinks" for these pollutants. These contaminants then can be released into the aquatic environment due to human activities such as dredging and due to biological and physical activity in rivers, lakes, and coastal waters.

An *Overview of Sediment Quality in the United States* was prepared for EPA in 1987.[105] This survey of 184 marine, estuarine, river, and lake or reservoir sediments found sediment contamination in all types of water bodies. The most frequent pollutants were heavy metals, PCBs and other organic chemicals, and pesticides. While all waters with past or present sources of pollution were contaminated to some degree, the most severe contamination was found in major harbors and heavily industrialized regions. Of the sites studied, those given the highest priority for cleanup were the Detroit River, Michigan; Baltimore Harbor, Maryland; Indiana Harbor, Indiana; Duwamish Waterway, Seattle, Washington; Michigan City Harbor, Indiana; and San Francisco Harbor, California. But the study authors warned that their findings may represent only a small part of the problem due to the small number of waters in which sediments have been tested and to large gaps in the pollutants measured in existing studies. While the study did not attempt a rigorous characterization of trends in sediment quality over the past twenty years, it concluded that the "historical record . . . shows that in-place contamination has increased rapidly through this century." Even with the report's limitations, it identified a wide range of human health and ecological impacts, including fishing bans or advisories, beach closures, fish kills, and sublethal effects on fish and other aquatic life.[106]

In 1989, the NRC's Committee on Contaminated Marine Sediments reached similar conclusions in its report on marine sedimentation contamination: "Sediment contamination is widespread throughout U.S. coastal waters and potentially far reaching in its environmental and public health significance."[107] Like EPA, the NRC found evidence of a wide variety of contaminants in sediments, including heavy metals, DDT, PCBs, and other organic chemicals. Contamination was noted "along all coasts of the contiguous United States, both in local 'hot spots' and distributed over large areas."[108]

Environmental Justice and Seafood Contamination

Seafood contamination appears to affect low-income populations disproportionately. Frequently, for geographic, economic, and sociological reasons, this translates to higher impacts on African Americans, Hispanic Americans, Asian Americans, Native Americans, and other subpopulations. These groups may consume much higher amounts of fish and shellfish caught from the most heavily polluted urban waters, for many of the same reasons as apply to swimming. While wealthier individuals can afford to travel to cleaner waters for recreational fishing, low-income city populations often have to choose between polluted fishing holes or none at all. Perhaps more important, wealthier populations can fish in polluted waters but not eat their catch, while in many cases low-income populations fish out of need—to put food on the table for themselves and their families.

A recent EPA report on environmental justice acknowledged that subsistence fishers who depend on their catch for food, as well as recreational fishers and some other groups, for cultural reasons, eat much more fish than the average person does.[109] A telephone survey regarding the consumption of fish from the Detroit River by a sample of Detroit residents confirmed this hypothesis.[110] The study compared minority and white consumption rates of four species fished from that river: white bass, walleye, sheepshead, and yellow perch (see Table 2.9).

Of the seventy respondents sampled, 15 percent said that they had caught and eaten fish from the Detroit River during the one-year period. "Of this sample, Whites tended to fish primarily for recreation, while minorities tended to fish for both recreation and food."[111] This means that Detroit River fish provide a supplemental source of protein for some people. "This adds a dimension of nutritional necessity to the policy debates over the need to protect water quality to permit safe access to fishery resources in urban areas."[112]

A more detailed follow-up study, using a mail survey to twenty-six hundred fishing license holders in Michigan, confirmed these results. It, too, showed higher average consumption by people of color than by white fishers (see Table 2.10).

Studies in other parts of the country, as well as national data, confirm these data. For example, a California survey of sport fishers showed that Asians eat more fish than other ethnic groups. A similar survey in Puget Sound found that Asians consumed a disproportionate amount of clams and the hepatopancreas of crabs, both of which concentrate toxic pollutants. And a national survey cited in EPA's manual for calculating exposure to toxic pollutants indicates that both Asians and African Americans consume more seafood than do Caucasians.[113]

Thus, while the public at large should be concerned about continued contamination of our fish and shellfish supplies, certain subpopulations face particularly acute health risks from polluted seafood in their local waters.

TABLE 2.9

Relationship between Race and Motivation for Fishing in the Detroit River		
MOTIVATION	**WHITE**	**NONWHITE**
Recreation	78.3%	40.0%
Food	0.0	2.0
Both	21.7	58.0
TOTAL	100.0	100.0

PERCENTAGE OF WHITES AND NONWHITES CATCHING AND EATING SELECTED SPECIES OF FISH FROM THE DETROIT RIVER

SPECIES	WHITE	NONWHITE
White bass	16.7%	52.2%
Walleye	30.0	35.6
Sheepshead	0.0	33.3
Yellow perch	16.7	35.6

Source: Patrick C. West, "Invitation to Poison? Detroit Minorities and Toxic Fish Consumption from the Detroit River," in *The Proceedings of the Michigan Conference on Race and the Incidence of Environmental Hazards,* ed. by Bunyan Bryant and Paul Mohai (1992).
Note: Percentage is of each category, not of total sample.

Aquatic Species and Ecosystems in Jeopardy

Noah's use of an ark to rescue the earth's wildlife from catastrophe bears ironic symbolism for the health of aquatic ecosystems, for aquatic and terrestrial species alike rely on our surface waters for habitat and sustenance.

The second dominant motivation for the Clean Water Act was the growing awareness that aquatic ecosystems were in serious jeopardy. By nearly a decade after Rachel Carson's ominous and prophetic warning in *Silent Spring,* it had become evident that water pollution and the global spread of persistent, highly toxic chemicals threatened more than human health—entire aquatic ecosystems were in jeopardy. Congress realized that a bandage would not cure this ecological disease. It insisted on no less than complete restoration of the bio-logical integrity of the nation's waters.

The most direct indication of biological integrity is the health of the species that inhabit an ecosystem. Two commenters defined biological integrity as the "ability of an aquatic ecosystem to support and maintain a balanced, integrat-ed, adaptive community of organisms having a species composition, diversity, and functional organization comparable to that of the natural habitats within a region."[114] Unfortunately, judged by the health of both aquatic and aquatic-dependent species, and of the habitats needed to support them, we are failing to meet this most fundamental goal of the Clean Water Act.

Disappearing Aquatic and Water-Dependent Species

Headlines in the 1980s and 1990s have captured the controversies between spotted owls and loggers and between red squirrels and astronomers in the struggle to preserve habitat for endangered species. Without minimizing the serious threats that these and other terrestrial species face, evidence shows, at least in North America, that fish and other species that inhabit or rely heavily on our aquatic ecosystems are actually faring much worse than their land-dwelling cousins. A shocking number of these species have already become extinct, and habitat loss, pollution, and other factors jeopardize many others.

TABLE 2.10

Average Fish Consumption by Race			
RACE	N	AVERAGE FISH CONSUMPTION	MAXIMUM VALUE* (G/PERSON/DAY)
Black	69	20.3	122.4
Native American	139	24.3	163.3
Other**	123	19.8	138.8
White	3,339	17.9	224.5
TOTAL	3,670	18.3	224.5

Source: Patrick C. West et al., "Minority Anglers and Toxic Fish Consumption," in *The Proceedings of the Michigan Conference on Race and the Incidence of Environmental Hazards*, ed. by Bunyan Bryant and Paul Mohai (1992).
* Maximum value is the highest fish consumption found for any individual in that age group.
** Includes Hispanics, mixed, and other.

Additional species, while not yet in direct danger of extinction, face plummet-ing populations and other problems.

The Nature Conservancy evaluated the status of selected animal groups in North America and found that aquatic species dominated the list of animals that are rare or threatened. While 13 percent of mammals, 11 percent of birds, and 14 percent of reptiles are identified as threatened, much larger propor-tions of aquatic species are at risk (see Figure 2.4).[115] Moreover, some non-aquatic species depend highly on aquatic ecosystems for survival.

A similar comparison based only on species listed as endangered or threat-ened as of 1990 by the FWS presents a far different picture. In this analysis, the distribution is far more evenly spread.[116] Apparently, FWS is predisposed to

FIGURE 2.4

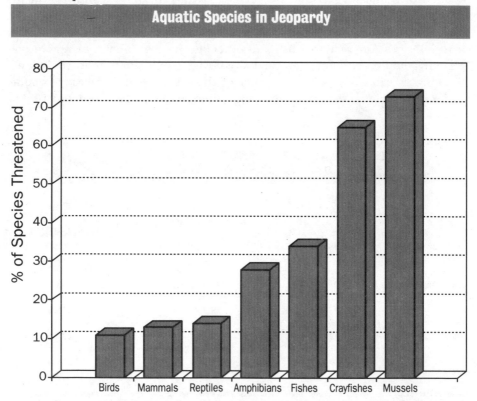

Aquatic Species in Jeopardy

Source: The Nature Conservancy, *Biodiversity Network News* 3, no. 3 (1990): 7.

devote its limited resources to terrestrial species, a tendency that led one biologist to ask, "Are conservationists fish bigots?"[117] The authors of a leading source on endangered fishes explained:

> When one thinks about endangered species, it is only natural to pay the most attention to endangered and threatened mammals such as whales, sea otters, wolves, and other furry, soft-eyed animals....
>
> Fishes, on the other hand, live in a world that is quite alien to ours. The general attitude toward preserving fish species that are either endangered or threatened has been one of "out of sight, out of mind." Most people find it difficult to identify with these cold-blooded, scaly vertebrates. But like the other species of wildlife that are losing their habitat through human exploitation, endangered and threatened fishes must be given a fair shot at survival...and equal attention.[118]

Even when the nation's experts convened in 1986 at the National Forum on Biodiversity, only two speakers addressed marine biodiversity, and no one spoke on freshwater aquatic biodiversity. "This is particularly disturbing when the state of aquatic diversity is rapidly approaching disaster."[119]

There are practical as well as ethical and ecological reasons for concern about threatened and endangered aquatic species. For example, understanding how the kidneys of desert pupfish can tolerate extreme salinity may help us redesign kidney dialysis units. And the Gila topminnow, a federally listed endangered species, is being used for cancer research in Germany.

Regardless of whether aquatic biodiversity has captured headlines, a more detailed analysis of which species are in danger, and why, is critical to a sound evaluation of how we are doing in our quest for biological integrity of the nation's aquatic ecosystems.

Trends in Aquatic Biodiversity

In 1964, Dr. Robert Rush Miller of the University of Michigan published the first list of endangered fishes. He characterized 38 species as endangered and 21 as "urgently threatened."[120] By 1979, the American Fisheries Society (AFS) had compiled a more comprehensive list of 251 North American fishes designated as endangered, threatened, or of special concern.[121]

Many of these species were profiled in a 1983 book called *Vanishing Fishes of North America,* by R. Dana Ono, James D. Williams, and Anne Wagner. Sewage and impoundments threatened darters in Maryland. Erosion and siltation, dredging and filling of wetlands, and industrial pollution endangered sturgeon in the Great Lakes. Septic seepage, fertilizer runoff, and erosion from development jeopardized killifish and shipjacks in the Southeast. Damming of the Colorado River system imperiled sixteen different fish species. Dace and darters in the Tennessee River system were devastated by the combined effects of damming and siltation and acid drainage from mining. During the century preceding the publication of *Vanishing Fishes of North America,* at least fourteen species and seven subspecies of fish became extinct in North America (six killifish, five minnows, four trouts, three suckers, and one each of perch, sculpin, and lamprey).[122]

The following excerpts from *Vanishing Fishes of North America* described the condition of fishes throughout the country:[123]

Atlantic and Gulf Coast Freshwater Fishes. "All of [these fishes] have felt the negative impact of Man's encroachment on their aquatic habitats.... Though these eight species of fish are separated by hundreds of miles in some cases, all face similar threats to their very survival."

Great Lakes Fishes. "The ciscoes of the Great Lakes probably represent the most significantly endangered fishery, the most significantly endangered fish populations, and the most significantly endangered combination of fish species in the freshwaters of the United States."

"The eight species of ciscoes . . . once were the dominant commercial fish of the Great Lakes, and the dominant forage fish for lake trout and other piscivores. Now the ciscoes are virtually extinct in Lakes Erie and Ontario, are restricted largely to Georgian Bay in Lake Huron, and are rapidly diminishing in abundance in Lakes Michigan and Superior." (From a report released on August 2, 1974, by the FWS Great Lakes Fishery Lab, Ann Arbor, Michigan.)

Pacific Coast Freshwater Fishes. "The native fishes of the Pacific Coast are silently disappearing from the low elevation regions where man's impact has had the greatest influence."

Colorado River System Fishes. "[I]f further habitat degradation occurs, the native Colorado River fish fauna may have the unique distinction of being the first major fauna to undergo mass extinction." (From a 1976 Department of the Interior document.)

Western Trouts. "Because of Man's alteration and overuse of the water resources in the western aquatic ecosystems, the native western trout populations have dwindled and some species have been pushed to the brink of extinction. Habitat for western native trout has been destroyed when water is diverted and stream flows reduced and when cattle overgraze adjacent rangeland and trample streambanks. Water quality problems resulting from timber harvesting and mining operations have reduced native trout populations in some streams."

Tennessee and Cumberland Fishes. "Today the Tennessee River system, with more than fifty major dams, is one of the most 'developed' rivers in the world. Remains of the widespread and diverse fish fauna and other aquatic organisms of the system are tucked away in the few miles of unaltered rivers that have escaped development."

Vanishing Fishes of North America identified, as the principal environmental causes of the demise of these species, the same environmental insults that prompted passage of the Clean Water Act. Apparently, however, implementation of the Clean Water Act has done little to reverse this trend. When AFS revisited its catalog of threatened and endangered fishes a decade later, in 1989, the situation had deteriorated severely. The 1989 list added 139 new taxa and removed 26, producing a total of 364 fishes that warrant protection due to rarity. These include 147 species characterized as of special concern, 114 as threatened, and 103 as endangered. Most of the species (254 out of 364) are in the United States.[124]

Nor should removal from the list be confused with improvement. Not a single species was removed from the list due to successful recovery efforts, while ten were dropped because they had become extinct. Of the species that changed categories, seven improved, twenty-four declined, and eighteen were reclassified for other reasons.

The AFS experts concluded that the factors that threaten most fishes had changed little since the 1979 classification: "Habitats continue to be degraded through human activities associated with agriculture, mining, industry and

TABLE 2.11

Fishes Identified as Extinct from 1979 to 1989	
Longjaw cisco	Thicktail chub
Deepwater cisco	Amistad gambusia
Backfin cisco	San Marcos gambusia
Alvord cutthroat trout	Blue pike
Independence Valley tui chub	Utah Lake sculpin

Source: Williams et al., "Fishes of North America Endangered, Threatened, or of Special Concern: 1989," *Fisheries* 14, no. 6 (1989): 13.

urban development, while harmful exotic species continue to be introduced and native fishes are transplanted beyond their natural ranges."[125] Overuse was a relatively minor factor and for only a few species. A companion article attempted to identify more precisely the principle causes of documented extinctions. Habitat alteration was the most frequent cause (one of the major factors in 73 percent of the cases), followed by introduced species (68 percent), chemical pollution (38 percent), hybridization (38 percent), and over-harvesting (15 percent).[126] These authors predicted an increased rate of extinction if current trends continue.

The most recent AFS survey also identified the number of endangered fish species in each of the fifty states. The figures range from a low of one jeopardized species in Alaska to highs of forty in Tennessee, forty-two in California, and forty-three in Nevada. Once again, this information suggests serious underreporting by many states of the percentage of waters that fail to meet the fishable waters goal of the Clean Water Act.

Remember that fishable is shorthand for "the protection and propagation of fish, shellfish and wildlife" and that the legislative history of the 1972 act indicates that *biological integrity* means a balanced, indigenous population of fish and wildlife.[127] Clearly, a primary goal is to prevent the extinction of (or other serious declines in) native fish populations. By definition, waters with endangered, threatened, or jeopardized fish species do not meet the fishable goal of the act. Yet, according to the 1990 National Water Quality Inventory, many states with large numbers of imperiled fish species report a remarkably small percentage of waters that are not fishable (see Table 2.12).

Certainly there is no direct numerical correlation between these two measures. A state may have a large number of imperiled species that reside in a small percentage of their waters, which may be the waters identified as degraded. And some species are listed as endangered because they are exotic or because of other impacts not necessarily relevant to the Clean Water Act.

On a national basis, however, one would expect states with many imperiled species to report a significant percentage of unfishable waters. It is difficult to explain how Georgia, Hawaii, Missouri, New York, Texas, Utah, and Wyoming each have a large number of imperiled fish species but report that 1 percent or less—and in some cases none—of their waters fail to meet the act's fishable goal. This may reflect the fact that some states view the fishable goal only in terms of immediate human health concerns. As discussed earlier, however, the law requires "protection and propagation" of balanced, indigenous populations of fish, shellfish, and wildlife. Clearly the 305(b) reporting system does not identify all waters that are jeopardized by chemical contamination, habitat loss, or other forms of ecological impairment.

TABLE 2.12

Imperiled Fish Species and Impaired Waters

STATE	NUMBER OF IMPERILED SPECIES	PERCENTAGE OF WATERS REPORTED AS NOT FISHABLE
California	42	7.70
Colorado	9	2.30
Georgia	20	0.20
Hawaii	4	0.00
Illinois	12	4.20
Kentucky	18	7.70
Louisiana	11	5.50
Michigan	7	2.00
Mississippi	14	2.90
Missouri	14	0.10
Nevada	43	11.70
New Mexico	20	8.50
New York	10	1.00
North Carolina	21	2.50
Oklahoma	10	9.50
Oregon	25	5.60
Tennessee	40	2.00
Texas	23	1.00
Utah	11	0.00
Virginia	21	4.40
Wyoming	8	0.03

Source: EPA, National Water Quality Inventory (1990); Williams et al., "Fishes of North America Endangered, Threatened, or of Special Concern," Fisheries 14, no. 6 (1989).

But fishes are not the only category of aquatic and aquatic-dependant species in jeopardy. The current FWS list of threatened and endangered species includes, in addition to ninety fishes, thirteen snails, forty-two clams and mussels, and ten aquatic crustaceans.[128] Many other species on the list rely heavily on aquatic ecosystems. The mammals, for example, include marine mammals such as the Florida manatee, stellar sea lion, and southern sea otter; wetlands or beach-dwelling species such as beach mice, voles, and shrews; and the Florida panther, whose habitat in the Everglades is facing increasing pressure from development. Currently listed bird species also include waterfowl and other species that use wetlands and other waters for food, nesting, staging, and other critical habitats. These include the whooping crane, the Mississippi sandhill crane, several species of ducks and geese, the Everglades snail kite, the brown pelican, the piping plover, the wood stork, and others.

Most experts agree, moreover, that currently listed species reflect only the tip of the iceberg. For each listed species, there are many others on the FWS candidate list,[129] including 70 reptiles; 63 amphibians; 172 fishes; clams and mussels; snails; insects, such as mayflies, dragonflies, damselflies, stone flies, caddis flies (whose role in aquatic ecosystems is well known to fly-fishers); and crustaceans, such as shrimp and crayfish. Again, even groups of predominantly "terrestrial" species include many that rely heavily on aquatic ecosystems for survival: mammals (three species of beaver, southwestern otter, beach vole, Florida long-tailed weasel, Everglades mink, Florida black bear); and birds (marbled murrelet, snowy plover, mountain plover, black tern, trumpeter swan, fulvous whistling duck, reddish egret, harlequin duck, western least bittern, black rail, band-rumped storm petrel, white-faced ibis, mangrove clapper rail, elegant tern, and common tern [Great Lakes population]).

Another indication of the health of species that rely on aquatic habitats is the percentage of threatened and endangered species in the coastal zone. According to biologists at the Nature Conservancy, 80 species and subspecies considered rare, imperiled, or critically imperiled are found only at elevations below 10 feet above sea level along the coasts of the contiguous United States. These include 46 plants, 14 mammals, 6 birds, 9 reptiles and amphibians, 3 fish, and 1 insect. The situation is even more acute for species that use the coastal fringe, often for critical portions of their life cycle, but are not restricted to this zone. For example, 73 species in Maryland are found only below the 10-foot contour, but as many as 122 species use this zone. In California, only 2 threatened species use the 10-foot zone exclusively, but species that use this zone heavily include the peregrine falcon, California brown pelican, light-footed clapper rail, California black rail, California least tern, Belding's savannah sparrow, and San Francisco garter snake.

Moreover, the implications of population declines transcend the status of individual species. Just as U.S. roadways have been transformed in recent years from ribbons through colorful regional diversity to multi-lane highways through monotonous patterns of cookie-cutter malls and fast-food franchises, the biota in our rivers, too, is becoming increasingly homogenized. One study, for example, showed that 67 percent of the fish species in the Illinois River and 44 percent from the Maumee River (in Ohio, Michigan, and Indiana) have disappeared or become less abundant.[130] While these species may not be endangered overall, the diversity of species in each river system has declined severely. With these losses, species that are more tolerant of human impacts increase, changing the ecological structure of the communities.

Unfortunately, as shocking as these statistics are, "extensive biological inventories of freshwaters have never been launched and subsequently the degree of biological impoverishment is largely unknown."[131] Given the extent of habitat losses, many more species have probably disappeared or are in serious jeopardy.

Trends in Aquatic and Other Water-Dependent Populations

Even where aquatic species and other species that depend heavily on aquatic ecosystems are not in immediate jeopardy of extinction, evidence shows that population levels are declining. Some of these species are valuable economic resources and sources of food; others are equally important for their ecological and aesthetic value.

America's coasts and some inland waters continue to provide a tremendous bounty of fish and shellfish, with vital economic and nutritional value to the nation. Indeed, many seafood populations are on the rise. These include the American lobster, landings of which have risen by more than half over the past twenty years. But other indicators are more ominous:

● Between 1970 and 1989, harvest of oysters dropped by 44 percent and landings of spiny lobster declined by 34 percent.

● Commercial landings of striped bass have declined continuously since 1973, with a fall of 92 percent since 1982.

● Between 1983 and 1989, landings of bay scallops fell by 88 percent. Scallop landings dropped by 50 percent from 1975 to 1985, with catch per unit effort in 1985 reaching historical lows.[132]

On a regional basis, losses can be devastating:[133]

● Between the mid-1960s and mid-1980s, Chesapeake Bay fish landings declined dramatically: hickory shad were down 96 percent, alewife and blueback herring down 92 percent, striped bass down 70 percent, and American shad down 66 percent.

● Columbia River basin salmon and steelhead have declined an estimated 75 to 84 percent from an estimated 10 million to 16 million fish historically to 2.5 million fish by 1991. Approximately 70 percent of those that remain are produced in hatcheries as mitigation for dam effects, as more than 55 percent of the Columbia basin has been blocked by dams.

● Sacramento River winter-run chinook salmon have declined 99 percent in the past twenty years, and they are now listed as threatened.

● English sole landings from Puget Sound have declined in recent years from a high of 2.4 million pounds to a low in 1987 of 0.7 million pounds.

● The commercial carp fishery in the Illinois River has virtually disappeared, dropping from more than 15 million pounds in 1908 to 4 million pounds in 1950 and 213,000 pounds in 1973.[134]

A 1991 National Fish and Wildlife Foundation assessment gives a somewhat more comprehensive view of the state of our marine fisheries, which provide

TABLE 2.13

Trends in Commercially Important Marine Species

NUMBER OF SPECIES OR GROUPS

REGION	LOW/ DECREASING	INCREASING/ HIGH/STABLE	UNKNOWN/ FLUCTUATING
North Pacific	11	9	6
Pacific	4	2	3
Western Pacific	3	4	2
Caribbean	2	—	—
Gulf of Mexico	5	4	1
South Atlantic	6	1	—
Mid-Atlantic	0	6	1
New England	2	2	—
TOTAL	33	19	14

Source: National Fish and Wildlife Foundation, National Marine Fisheries Service, FY 1992 Wildlife and Fisheries Assessment (April 1991), App. B.

the vast majority of our commercial catch of natural seafood (see Table 2.13). This assessment shows that far more species of important commercial seafood are declining than are increasing or remaining stable.

Again, these trends are not limited to species that spend all of their time in the water but include other water-dependant species as well. The most frequently cited example is waterfowl populations, which in general have plummeted over the past quarter century. According to the CEQ's 1989 national assessment of environmental trends, duck breeding populations in North America dropped continually from 1955 through 1985.

● Mallards reached a record low of 5.5 million in 1985 (from about 13 million in 1958); causes include habitat destruction from draining and filling wetlands and lakes and farming marginal lands adjacent to water bodies, and use of chemicals on adjacent farms.

● The black duck population has declined by 60 percent since 1955; the biggest causes are habitat alteration and hybridization with mallards.

● Northern pintail declined from 1955 to a record low of 2.9 million in 1985; causes include reductions in wetlands and quality of habitat in the breeding range.

● Canvasback ducks have exhibited dramatic reductions since 1980; losses are directly related to degradation of waterfowl habitat in selected areas, decline in preferred food, submerged aquatic vegetation due to declining water quality, and increased siltation.[135]

More recent data suggest that this trend has not been reversed. The ten species with more than 97 percent of North America's breeding populations showed declines of 32 percent from 1972 to 1992, the period when the Clean Water Act was in place to protect aquatic habitat.[136]

And waterfowl are not the only water-dependent birds showing serious declines. Data from the FWS Breeding Bird Survey, which has recorded flight records since 1966, show that a significant number of water-dependent species have declining population trends (see Table 2.14).

Fish Kills and Aquatic Toxicity

Even where species or populations are not in immediate danger of extinction or precipitous population declines, ongoing pollution is having serious impacts on fish and other aquatic species. Pollution continues to cause hundreds of fish kills every year, affecting millions of fish. And even where fish do not die, there is increasing evidence of serious impacts of water pollutants on aquatic species, such as tumors, growth impairment, infertility, and behavioral changes.

TABLE 2.14

Water-Dependent Birds with Declining Population Trends (Not Including Waterfowl)

2 of 4 cormorants (Brandt's, pelagic)
....

5 of 19 herons, ibis, storks (American bittern, great white heron, little blue heron, green-backed heron, black-crowned night heron)
....

3 of 9 rails, gallinules, coots (Virginia rail, sora, limpkin)
....

8 of 16 shorebirds (mountain plover, American black oystercatcher, American avocet, lesser yellowlegs, spotted sandpiper, long-billed curlew, American woodcock, Wilson's phalarope)
....

7 of 11 gulls (Franklin's gull, Bonaparte's gull, Heermann's gull, California gull, herring gull, western gull, great black-backed gull)
....

6 of 9 terns (gull-billed tern, royal tern, common tern, least tern, black tern, black skimmer)
....

Others (American white pelican, anhinga, pigeon guillemot, belted kingfisher, green kingfisher)
....

Source: FWS, Patuxent Wildlife Research Center, Computer Printout of Breeding Bird Survey Trends 1966–1991.[137]

Fish Kills

The U.S. Public Health Service began reporting pollution-caused fish kills in 1960. On passage of the Clean Water Act in 1972, this function was transferred to EPA. Summary reports were prepared by EPA for the periods 1961 to 1975 and 1977 to 1987,[138] and data for 1988–89 are available in the 1990 *National Water Quality Inventory.* Because of this reporting history, fish kill data provide one of the best long-term indicators of the success of water pollution control efforts. As EPA notes, "[o]ne obvious and important indicator of water quality problems is the occurrence of fish kills caused by pollution."[139]

Unfortunately, these data show little if any progress in stopping fish kills caused by pollution. From 1972 to 1989, EPA estimates, at least 428 million fish were killed in more than ten thousand incidents. These data underrepresent the number of fish kills for several reasons: (1) reporting is voluntary and some states do not participate; (2) some fish kills go unnoticed, and others are not included in reports because causes are not known; (3) data are lost where investigation is delayed, because fish are washed away and pollution is diluted; and (4) the number of fish estimated to be killed often is conservative—in some cases, as many as 80 percent of the dead fish cannot be counted because of turbid water or settling to the bottom. Problems also exist with data consistency because of wide variation in how states complete reports and because many reports are incomplete.[140]

At first blush, the number of fish kills appears to have declined over time. Until 1981, states generally reported between 700 and 850 fish kills per year; this level dropped to 500 or fewer for much of the 1980s. But these numbers are skewed by extreme variations in the numbers of states reporting fish kills during certain years. Reporting dropped from all states in the early 1970s to an average of thirty-six states from 1977 to 1985, with a low of twenty-four in 1986 and with forty-two reporting in 1988–89. Adjusted to reflect these variations in the number of states reporting, the trends show an increase from roughly 700 to 750 incidents per year in the early 1970s to between 800 and 1,000 incidents per year in the late 1970s and 1980s. While this increase may reflect better detection and reporting by states in the later years, it also indicates that large numbers of fish kills continue two decades after the Clean Water Act was passed.

The total numbers of fish killed each year have fluctuated more widely because total kills in a given year may be skewed by catastrophic single events. For example, about 17 million fish were killed in 1972. This number jumped to almost 39 million in 1973 and skyrocketed to more than 119 million in 1974 but dropped back to just over 16 million in 1975. In general, these levels appear to have dropped in recent years. Breaking down the data into three six-year periods, states reported an average of more than 40 million fish killed per year from 1972 to 1977, compared to roughly 28 million per year from 1978 to 1983 and 8.5 million per year from 1984 to 1989. With reported fish killed normalized to reflect differences in state reporting, this trend is somewhat less pronounced but still evident (see Table 2.15).

A second source of data indicating trends in fish kills is available from NOAA, which tracks kills in coastal waters. From 1980 to 1989, more than 3,650 fish kills were reported in 533 coastal and near-coastal counties in twenty-two states. More than 407 million fish were killed in these episodes. The number of events reported was highest in 1986 (519), and the greatest number of fish killed (138 million) was in 1980.

It is somewhat easier to detect trends in the NOAA data, because the number of states reporting each year has not varied widely. In general, the trends parallel those shown by EPA: The total number of reported fish kill incidents increased during the 1980s, but the average numbers of fish killed per year have dropped (see Table 2.16). This trend is less apparent, although still present, if the numbers are adjusted to eliminate the largest kill each year, which

TABLE 2.15

Inland Fish Kills Caused by Pollution

YEAR	STATES REPORTING	# OF REPORTS TO EPA	ESTIMATED INCIDENTS FOR 50 STATES	TOTAL FISH REPORTED KILLED (MILLIONS)	TOTAL KILLED (ESTIMATED FOR 50 STATES) (MILLIONS)
1972	50	760	760	17.7	17.7
1973	50	754	754	37.8	37.8
1974	50	721	721	119.1	119.1
1975	50	624	624	16.1	16.1
1976	—	—	—	—	—
1977	41	503	613	16.5	20.1
1978	42	686	817	74.7	88.9
1979	47	777	827	8.1	8.6
1980	43	801	931	29.9	34.8
1981	43	845	983	50.2	58.4
1982	31	499	805	2.7	4.4
1983	22	501	1,139	4.2	9.5
1984	29	507	874	2.4	4.1
1985	25	484	968	10.4	20.8
1986	29	514	886	2.1	3.7
1987	24	476	992	9.8	20.5
1988	42	683	813	13.0	15.5
1989	42	683	813	13.0	15.5
TOTAL		10,818	14,320	427.8	494.5

Sources: EPA, *Fish Kills Caused by Pollution, 1977–1987*; EPA, *Fish Kills Caused by Pollution, Fifteen-Year Summary 1961–1975*; 1988–90 figures from EPA, *National Water Quality Inventory* (1990), 305(b) Report to Congress (March 1992).

tends to skew the data high for years with a single catastrophic event, such as 1980 and 1981.

We estimate that at least 1.35 billion fish have been killed in inland and coastal waters together since the Clean Water Act was passed.[141] These trends show that much more progress is needed to eliminate sources of pollution that generate major fish kills.

Aquatic Toxicity

Increasing evidence indicates that even extremely low, sublethal levels of toxic pollutants can cause a wide range of health effects to fish and wildlife. Particularly, pollutants that are persistent and bioaccumulative continue to be

TABLE 2.16

Trends in Coastal Fish Kills			
YEAR	# OF STATES REPORTING	# OF EVENTS	EST # OF FISH KILLED (MILLIONS)	LARGEST KILL (MILLIONS)
1980	21	279	138	50
1981	21	358	97	30
1982	16	283	12	2
1983	15	283	22	4
1984	17	263	41	22
1985	18	340	33	8
1986	20	519	24	2
1987	20	424	4	1
1988	19	464	32	18
1989	18	442	6	3

Source: J. A. Lowe et al., *Fish Kills in Coastal Waters, 1980–1989*, Strategic Environmental Assessments Division, Office of Ocean Resources Conservation and Assessment, National Ocean Service, NOAA (Rockville, Md.: 1991).

released into our aquatic environments at levels that are toxic to aquatic species and to birds, mammals, and other predators that consume contaminated fish.

The literature on aquatic toxicity is becoming so voluminous that it is impossible to characterize it fully here. One series of experiments in which rainbow trout were exposed to the most toxic form of dioxin (2,3,7,8-TCDD) illustrates the minute levels of exposure that can seriously affect aquatic species, however.[142] Exposure over twenty-eight days resulted in significant increased mortality and sublethal effects:

● Significant mortality was shown within fourteen days at just 176, 382, and 789 picograms per Liter (pg/L; a pg/L is equal to one part per quadrillion, or one molecule of dioxin for every quadrillion molecules of water; 1 quadrillion is 1,000,000,000,000,000).

● A significant decrease in growth occurred at 38 pg/L after twenty-eight days.

● Significant behavioral impairments were also shown (at the same levels), including lethargic swimming, feeding inhibition, and lack of response to external stimuli. These effects became progressively worse over time. Fish were "seriously stressed," with an abnormal head-up swimming position and confinement to the bottom of the aquarium, and effects were not reversed after twenty-eight days of exposure to clean water.

Much of the growing body of evidence about the effects of toxics on fish and wildlife comes from the Great Lakes and has been presented in useful summaries. For example, a 1991 report by the NWF and the Canadian Institute for Environmental Law and Policy summarized the effects of toxic contaminants on wildlife in the Great Lakes area:[143]

Population Declines and Reproductive Problems—Bald eagles in the Great Lakes area have lower reproductive rates than eagles living inland. There are virtually no minks left within a five mile radius of Lake Ontario. Fifteen kinds of birds, animals, and fish in the Great Lakes region have had reproductive problems and/or population declines since the 1950s.

Birth Defects—Twisted beaks and deformed eyes cause many young fish-eating cormorants and terns from Green Bay and Saginaw to die from starvation. Double-crested cormorants in four island colonies of Green Bay were born with bill defects 42 times more often than in colonies outside the Great Lakes region. Missing brains, missing eyes, internal organs located outside the body, and deformed feet and wings are other abnormalities found in Great Lakes wildlife. Birth defects occurred in almost 50 percent of the species studied.

Behavioral Changes—There is increasing evidence of behavioral changes that risk the survival of Great Lakes species—gulls ignore their eggs; terns leave their eggs at night, leaving them vulnerable to predators; young lake trout swim upside down. Six species of wildlife have shown serious documented behavioral changes.

Sexual Changes—Male herring gull chicks from Lake Ontario were found with female organs. Similar abnormalities were found in minks and lab animals. These changes are thought to be caused by the similarity in structure of PCBs, DDE, and other pesticides to female hormones.

Increased Susceptibility to Disease—Beluga whales, terns, and herring gulls have suffered a suppression of their immune systems.

Scientists feel that toxics are the cause of these abnormal health problems. They have discovered, for example, that

Female minks fed a 30% diet of Lake Michigan salmon produced no live young, while control mink fed West Coast salmon were not affected.

Levels of only one part per million of PCBs in mink livers are associated with total reproductive failure.

Rats fed Lake Ontario salmon easily became frustrated, anxious, and less active than those rats in a control group.

The health and behavioral problems are more associated with fish-eating birds and mammals whose diets consist of Great Lakes fish. This disruption of the food chain could deplete biological diversity and have unpredictable consequences for the ecosystem.[144]

Aside from the data on trends in toxic contaminants in fish and sediment presented earlier, little usable information is available to evaluate trends in aquatic toxicity since 1972. Such information is lacking largely because so much of our knowledge on aquatic toxicity is relatively recent. Nevertheless, information from other regions suggests that the Great Lakes findings are not isolated. For example:

● Exposure of adult English sole to contaminants (1) causes failure of egg development in some of the fish, (2) interferes with the timing of the spawning of those that do produce eggs, and (3) results in deformed young of those that do spawn on time. Almost 40 percent of the female English sole that the National Marine Fisheries Service (NMFS) tested from Eagle Harbor and Duwamish Waterway, both in Washington, failed to mature sexually. Sediments in both areas have high levels of aromatic hydrocarbons, which can result in disease and induce a variety of lesions leading eventually to development of cancerous tumors.

- Liver cancer (the most extreme lesion) has been found in 20 percent of English sole collected from two of the most contaminated areas of Puget Sound and in 15 percent of winter flounder samples from similarly affected areas of Boston Harbor.[145]
- Liver cancer and precancerous liver lesions have been found in 33 percent and 93 percent, respectively, of the killifish collected from a contaminated site in the Elizabeth River, Virginia. Virtually all adult gray trout (which feed by sight) collected from heavily polluted areas of that river have contaminant-induced eye cataracts.[146]

More broadly, the Smithsonian Institution's Registry of Tumors has documented that fish with serious contaminant-related abnormalities are generally found in those areas of the United States affected most by coastal pollution from about nineteen hundred major industrial and municipal dischargers.[147]

Lost of Damaged Aquatic Habitats

The serious declines in aquatic species discussed in the previous section result from multiple causes. At the top of the list is loss or destruction of aquatic habitat. We have dammed or otherwise altered most of our free-flowing rivers; more than half of our wetlands and riparian habitat has been filled, drained, or developed; and much of our prime spawning, rearing, and other fisheries habitat has been destroyed by siltation and other human activities.

Most experts agree that overall habitat loss is the single largest factor in the decline of aquatic species. "[A]s civilizations have advanced, so too have the ways by which humankind has purposefully or inadvertently harmed streams and rivers."[148] Accordingly, there is virtual unanimity among experts that single-issue and single-species solutions will not suffice. Declines in single species are just an indicator of more serious ecological problems. Instead, "conservationists should consider the protection of entire watersheds and ecosystems."[149]

There is no single source of information on loss or degradation of aquatic habitats since 1972. However, a significant amount of objective information exists on particular types of aquatic habitat. Some is quantitative and useful in detecting long-term trends. Some is more qualitative but confirms the sad reality that while some chemical pollutant loads may be declining, the overall integrity of our aquatic ecosystems is deteriorating dramatically.

Instream Fisheries Habitat

To help fill the gap in knowledge of the overall biological health of rivers and lakes, in 1982, EPA and FWS conducted the National Fisheries Survey, billed as the first statistically designed survey of the status of the nation's waters and fish communities. The survey requested detailed information on habitat conditions from professional aquatic biologists around the country, based on a large statistical cross section of waters. Unfortunately, despite its conclusion that much follow-up work was warranted,[150] the National Fisheries Survey was both the first and the last effort of its kind.

The conclusions of the survey were striking: 81 percent of the nation's waters, including 53.3 percent of all perennial waters, had fish communities adversely affected by a variety of factors.[151] (Perennial waters run continuously, as opposed to intermittent streams, which run only during wet periods of the year or only after sufficient precipitation.) Most telling were the survey's rankings of waters (see Table 2.17).

TABLE 2.17

Health of Inland Fisheries Habitat		
RANK	**ALL STREAMS**	**PERENNIAL STREAMS**
0	23.1%	3.1%
1	9.7%	5.2%
2	21.3%	17.3%
3–4	42.1%	40.1%
5	3.9%	3.8%
KEY		
0	no ability to support any fish population	
1	ability to support nonsport fish only	
2	minimal ability to support sport fish or species of special concern	
3–4	intermediate ability to support sport fish	
5	ability to support sport fish, species of special concern, or both at maximum level	

Source: EPA, FWS, *National Fisheries Survey* (1984).

All told, even for perennial streams, more than one out of four provided only minimal support or less for healthy fish populations. Less than 4 percent of waters were rated as completely healthy. Moreover, relative to five years earlier, the survey found the following trends: 91 percent of all waters in the survey kept the same rank, 4 percent improved, and 5 percent degraded. Respondents believed conditions would deteriorate severely over the next five years.[152]

To put these findings in perspective, the *National Water Quality Inventory* released the same year (1984) found that almost three-quarters of all rivers and streams were "fully supporting" designated uses. Remember that the Clean Water Act stipulates that full use support for aquatic life means protection and propagation of a balanced, indigenous population of fish and other aquatic species. This claim cannot be squared with the identification of only 4 percent of waters as healthy (or even 44 percent in the intermediate and healthy categories combined) by the National Fisheries Survey. Indeed, given the absence of any comprehensive effort to protect these fisheries habitats since 1984, and the predictions of most respondents that conditions would deteriorate in the absence of such efforts, the results are even harder to reconcile with claims in the 1990 *Water Quality Inventory* that more than 80 percent of all waters fully meet the fishable goal of the act, and less than 1 percent fail to meet this use at all.[153]

Wetlands

Wetlands provide important habitat for many threatened and endangered species. In fact, a recent report by the NWF concluded that "[a]lthough wetlands occupy less than 5 percent of the land area of the lower 48 states, 43 percent of all federally listed threatened and endangered species utilize wetlands at some point in their life cycle."[154] Virtually every category of plant and animal contains threatened and endangered species that rely heavily on wetlands, with fishes and plants leading the list.[155]

Thanks to monumental nationwide efforts by the FWS, better trend information is available for wetlands losses than for most other aquatic ecosystems. A 1991 report to Congress on the status and trends of wetlands summarized this information from the mid-1970s to the mid-1980s.[156] The availability of this information is extremely fortunate, because wetlands are among our most important aquatic ecosystems, yet they are under perhaps the most severe threats and have been the victim of the worst complete-habitat loss, as opposed to contamination or partial degradation.

This analysis is shocking in its revelation of overall wetlands losses throughout the nation's history. At the time of European colonization, the coterminous United States had an estimated 221 million acres of wetlands. More than half of this acreage has been lost through draining, dredging, filling, levying, and flooding. Twenty-two states have lost 50 percent or more of their original wetlands acreage, and ten have lost more than 70 percent.[157]

When the Clean Water Act began to regulate wetlands in the mid-1970s, an estimated 105.9 million acres of wetlands remained. By the mid-1980s, only 103.3 million acres remained, with a total loss of 2.6 million acres, or an average of 260,000 acres per year.[158] Estuarine wetlands have declined by about 1 percent—mostly in Gulf Coast states—in most cases due to conversion to open salt water. Inland vegetated wetlands have decreased by nearly 2.5 million acres, with the largest losses in forested wetlands, primarily in the South.[159]

Especially given the magnitude of historical losses, a loss of more than .25 million acres a year remains unacceptable and is fundamentally inconsistent with the objective of restoring and maintaining the integrity of the nation's waters. Unlike many forms of pollution and other partial habitat changes, dredging and filling of wetlands results in complete or virtually complete loss of aquatic habitat and other wetlands functions. In most cases, these losses are permanent.

At the same time, the study is at least somewhat promising from the perspective of recent trends. The rate of wetlands loss has slowed by about half since the wetlands protection provisions of the act have been in place. If this trend continues and wetlands protection efforts are expanded and strengthened, perhaps we can reverse the tide and begin to restore rather than degrade the nation's wetlands resources.

The FWS report also shows that the causes of wetlands loss have shifted. Between the 1950s and the mid-1970s, nearly all conversions were due to agriculture. From the mid-1970s to the mid-1980s, losses were split more evenly between agricultural (54 percent) and other (41 percent) categories. Urban land development was responsible for an almost 60,000-acre net loss in forested wetlands and for almost 60,000 acres of losses of other vegetated wetlands.[160]

These figures apply only to the complete loss of wetlands acreage, however. They do not address pollution or more subtle forms of damage to wetlands, which, like pollution and degradation of other aquatic ecosystems, can result in the significant loss of natural and other values. Few states have standards to address wetlands uses or specific criteria to determine whether those uses are protected. In the 1990 *National Water Quality Inventory*, for example, only five states (California, Hawaii, Iowa, Kansas, and Nevada) reported on use impairment for wetlands. The information in these reports, however, is disturbing. The five states combined estimated that only about half of their wetlands fully supported designated uses, with almost 6 percent threatened, 26 percent partially supporting, and more than 17 percent not supporting wetlands uses.[161] While the paucity of data makes extrapolations to other states virtually impossible, this information suggests strongly that we have lost far more than half of the nation's wetlands in terms of the functions they perform, as opposed to raw reductions in acreage.

Floodplains and Riparian Habitat

Like wetlands, floodplains (which in many cases overlap with wetlands)[162] and riparian (streamside) habitat provide essential hydrological and ecological support functions for the adjacent rivers, streams, and lakes. Floodplains have high biological productivity due to exchanges of water and materials between river and floodplain, with a highly diverse biota adapted to floodplain habitat and resources.[163] They are generally more biologically diverse than adjacent uplands and encompass a broader range of moisture and soil conditions.

Healthy riparian systems provide a range of ecological values, such as community structure for raptors, safe passage corridors to water for mammals, habitat for amphibians, and cover and nutrients for fish. These corridors are particularly important in the arid West, where they can provide habitat for up to 80 percent of wildlife species.[164] Less than 1 percent of the western landscape of the U.S. is covered by riparian vegetation; however, this vegetation provides habitat for more species of breeding birds than surrounding uplands. For example, 82 percent of all species that annually breed in northern Colorado occur in riparian vegetation, and 51 percent of all species in southwestern states are completely dependent upon this vegetation type. Loss of the riparian component in the southwestern states could potentially result in the loss of 78 (47 percent) of the 166 bird species that breed in the region. Studies indicate that the San Juan and Gila valleys, New Mexico, support 16%–17% of the entire breeding bird population of temperate North America, and similar relationships have been observed in the eastern United States.[165]

Estimates of the amount of floodplain and riparian habitat vary widely, depending on the definition used. The U.S. Water Resources Council estimated in 1977 that almost 180 million acres are within the hundred-year floodplain in the United States (the riparian zone expected to be inundated by floodwaters an average of once every century). The Soil Conservation Service (SCS) estimates that there are about 16 million acres of riparian land along streams, canals, lakes, reservoirs, and tidal shores. Bottomland hardwood forests account for another 52 million riparian acres.[166]

Estimates of loss of floodplain and riparian habitat also vary. But while they differ in detail, all lead to the conclusion that a large percentage of the original riparian habitat in the United States has been lost, and a large percentage continues to be lost. A detailed 1992 assessment of floodplain management in the United States provides useful perspectives:

● By the late 1970s, an estimated 3.5 million to 5.5 million acres of floodplain had been developed for urban use, including more than six thousand communities with populations of twenty-five hundred or more. Annual growth in these floodplains during the 1970s was between 1.5 percent and 2.5 percent, roughly twice that of the country as a whole.

● Out of 75 million to 100 million acres of indigenous, woody riparian habitat, less than half (about 35 million acres) remain in nearly natural condition. The rest have been inundated, channelized, dammed, riprapped, farmed, overgrazed, or altered by other land uses.

● The Corps estimates that there are 574,500 miles of stream bank with erosion problems in the United States, 142,100 of which are characterized as "serious."[167]

These figures alone do not provide a clear picture of changes since the Clean Water Act was passed in 1972. But given that implementation of the act has focused on regulation of point source discharges of pollution, it probably has had little impact on the rapid pace at which we have destroyed riparian habitat.

We know that much of the damage was already done by the time the act was passed. By 1973, the channels of at least 200,000 miles of waterways in the United States (approximately 1 of 5 stream miles) had been modified. This would equal more than half the total length of large warm-water streams where channel alterations are most prevalent. This estimate is consistent with the findings of the U.S. Heritage Conservation and Recreation Service, which determined that approximately 60 percent of the miles of large streams and rivers in the United States were unsuitable for inclusion in the National Wild and Scenic Rivers System, because of water resource or other cultural developments within the riparian corridors. Large reservoirs (greater than 500 surface acres), which presently cover about 10 million acres, probably inundated at least 15,000 miles of stream.[168]

Evidence exists that these trends have continued since the Clean Water Act was passed. For example, the amount of SCS channel work constructed or under contract totaled nearly 11,000 miles by 1980, translating to an average loss of approximately 300 miles of natural stream habitat per year since 1972 from this source alone. By the end of the 1970s, multiple sources confirmed that at least half of this country's original riparian habitat had been destroyed, with devastating impacts on the health of our aquatic ecosystems. According to one 1979 estimate, only 23 million out of 67 million acres of the four predominant riparian vegetation types remained in the United States, a loss of 66 percent. In 1981, FWS biologists calculated that only 26 million out of almost 56 million acres of riparian ecosystems remained in the contiguous states and Hawaii.[169]

Alteration of Stream Flow

Dams, diversions, and other structures that alter the natural flow of rivers cause the loss of a significant amount of natural riparian habitat. Dams and reservoirs create many unfavorable environmental changes. For example, they inundate wetlands and destroy valuable fish and wildlife habitat, trap sediment, cause downstream river erosion and channel alterations, degrade water quality, and reduce or alter the timing of flows available to downstream ecosystems. The effects of water projects have filled entire volumes and cannot possibly be addressed adequately here. But several key facts underscore the contribution of such projects to lost or damaged aquatic habitat:

● The number of dams in the United States increased dramatically between 1935 and 1986, with no significant decline in the rate of growth since 1972. By the mid-1980s, there were approximately 2,654 large reservoirs and controlled lakes (each with a capacity greater than or equal to 5,000 acre-feet), more than 50,000 smaller facilities (50 to 5,000 acre-feet), and about 2 million smaller farm ponds.[170]

● By 1980, large reservoirs alone (greater than 500 acres) probably had inundated at least 15,000 miles of streams.[171]

● More than 68,000 nonfederal dams have altered or destroyed an additional tens or hundreds of thousands of miles of riparian habitat.[172]

While it is difficult to quantify on a national scale the magnitude of habitat loss due to dams and other water projects, available information suggests that losses have been devastating. In order to identify candidates for the Wild and Scenic Rivers System, an inventory of the nation's rivers (the National Rivers Inventory [NRI]) was conducted during the 1970s and 1980s (until 1980 by the Heritage Conservation and Recreation Service and then by the National Park Service). This survey concluded that out of approximately 3.2 million miles of rivers and streams in the United States, less than 2 percent were of sufficiently high ecological quality to be selected for the inventory.[173] The majority of river segments were excluded because of flow and habitat alteration for hydropower or navigation.

An updated and more comprehensive analysis of the NRI database by Dr. Arthur C. Benke of the University of Alabama confirmed the small number of stream segments left in the contiguous states that have not been significantly altered by human development (and even this analysis was restricted primarily to the impacts of hydroelectric development). All but one river longer than 620 miles, or 1,000 kilometers (the Yellowstone River in Montana), have been severely altered, with most converted to a string of artificial reservoirs.[174] Of the more than one hundred rivers in the database greater than 124 miles (200 kilometers) in length, only forty-two were still free flowing.[175] Only eight of these are subject to permanent federal protection against alteration for hydropower or other purposes.[176] And Dr. Benke noted that even these streams, while free of major obstructions or channel alterations, "are by no means pristine. Many represent the most natural segments on river systems with otherwise severe modifications."[177]

Unfortunately, Dr. Benke's analysis shows that the Clean Water Act and other environmental laws have not curbed the rate at which water projects alter natural aquatic habitat and thus has done little to protect the biological integrity of free-flowing rivers and streams. Total hydroelectric-generating capacity in the United States nearly doubled between 1970 and 1990 and is projected to grow at only a slightly slower rate through the turn of the century.[178] Worst of all, most of the new sites slated for hydro development in the future will generate only trivial amounts of power. Since most sites suitable for large dams have already been used, a large number of smaller projects are planned instead. According to the Federal Energy Regulatory Commission, which licenses hydro facilities, the eleven hundred-plus small projects planned as of 1988 would increase by more than half the total number of dam sites in the country, in return for a tiny (2 percent) increase in hydroelectric power and only 0.3 percent of total electricity capacity in the country.[179] Dr. Benke concludes, "With so little to gain in the way of energy, any project affecting streams of high natural value . . . would appear difficult to justify."[180]

Coastal Habitat

While aquatic habitats are suffering throughout the country, the cumulative impacts of upstream degradation converge on coastal ecosystems, combining with ever-increasing pressures for economic and population growth in coastal regions. Addressing the Fifty-seventh North American Wildlife and Natural Resources Conference in 1991, a chief NOAA scientist said:

> The evidence of the decline in the environmental quality of our estuaries and coastal waters is accumulating steadily. The toll of nearly four centuries of human activity becomes more and more clear as our coastal productivity declines, as habitats disappear, and as our monitoring systems reveal other problems. . . . The continuing damage to coastal resources

from pollution, development, and natural forces raises serious doubts about the ability of our estuaries, bays, and near coastal waters to survive these stresses. If we fail to act and if current trends continue unabated, what is now a serious, widespread collection of problems may coalesce into a national crisis by early in the next century.[181]

In addition to toxic chemicals and excess nutrients, NOAA cited habitat degradation as the principal cause of this decline. Sources of coastal habitat degradation include freshwater flow alteration and diversion, wetland conversion, erosion and habitat loss from land development, dam construction, navigation channel construction, port development, energy production, logging, agriculture, and other resource-consumptive uses in adjoining watersheds. "Because these varied demands can adversely affect the ability of natural systems to support aquatic life and maintain their ecological integrity, competition and conflict over the fate of inshore habitats have risen with the accommodation of increasing coastal and inland development."[182]

The clearest statistical indicator of this trend is the incessant growth pressure in coastal regions. Coastal counties are growing at four times the national average. By the year 2010, according to estimates, about 45 percent of the U.S. population will live within 50 miles of the coast.[183] Population trends alone, of course, do not necessarily translate into lost or degraded habitat. But available evidence shows that many of the habitat effects are concentrated in the coastal zone. Following are some examples.

Overall Habitat Losses. Before World War II, only one-quarter of all barrier islands were subject to development pressure; by 1980, this had increased to 70 percent. Half of the Chesapeake Bay's wetlands, 90 percent of its sea grass meadows (prime nursery habitat, with 65 percent between 1971 and 1979), and 40 percent of its forested areas have been destroyed. More than 90 percent of California's salmon spawning habitat has been lost due to extensive federal and state water projects in the Central Valley.

Freshwater Flow Alterations. Alteration of flows by dams or diversion as well as by land-use practices (such as logging) may be the single most important factor influencing the health of many estuarine ecosystems. A 1981 national symposium on the effects of freshwater diversions concluded that, "based on worldwide experience, no more than 25–30% of the historical river flow to an estuary can be diverted without disastrous ecological consequences to the receiving estuary." Yet tributary flows in the Chesapeake Bay have been reduced by 40 percent; Texas estuaries have lost nearly 90 percent of their historical inflows due to upstream diversions; by 1980, more than 62 percent of the annual historical freshwater inflows to San Francisco Bay had been diverted; and planned diversions will increase annual loss of fresh water to 71 percent by the year 2000.

Wetland Losses. The Southeast has more than three hundred estuaries containing an estimated 17.2 million acres of coastal marsh. Commercial fishery landings along Southeast Atlantic and Gulf coasts have decreased by 42 percent since 1982. These decreases were coincidental with extensive regional losses of coastal habitats. For example, Louisiana's coastal wetland losses between 1974 and 1983 are estimated as more than 30 square miles per year. Galveston Bay, Texas, lost an estimated 95 percent of its former sea grass meadows and 16 percent of its emergent marsh between 1959 and 1979. By 1981, coastal development eliminated an estimated 81 percent of Tampa Bay, Florida's extensive sea grass, and 44 percent of its emergent marsh and mangrove habitats. More than 91 percent of California's coastal wetlands have been lost; San Francisco Bay wetlands have declined by 85 percent.[184]

A 1992 NOAA report documenting the tremendous amount of construction along U.S. coastlines in the past two decades confirmed the principal cause of these coastal habitat losses. During 1970–89, an average of 384,000 new single-family houses were approved for construction in coastal counties—nearly half of the national total in just 11 percent of our land area. For multifamily units, more than half of all construction was in coastal counties. Similarly high construction rates were evident in the commercial and industrial sectors. During this period, coastal counties approved the building of almost 300,000 retail buildings, more than 152,000 office buildings, more than 172,000 industrial buildings, and more than 19,000 hotels.[185]

If these trends continue, clearly we stand to lose even more of our precious remaining coastal habitat.

Conclusion

In this chapter we surveyed a broad range of indicators of how well the goals of the Clean Water Act have been met over the past twenty years. This analysis shows that significant progress has been made in some areas: Total releases of toxic and other pollutants are down, due to investments in new pollution controls and other measures by public sewage plants and industries; many waters that were barren of fish once again support commercial and recreational fisheries; and some waters that were badly contaminated are now safe—or at least safer—for fishing and swimming.

But serious problems remain even after two decades of dedicated effort. We continue to release large quantities of pollutants into our rivers, lakes, and coastal waters, resulting in ongoing risks to human health and environmental quality. Beaches remain closed or, worse yet, polluted without warning; fisheries are still closed or subject to consumption advisories; and public drinking water supplies remain contaminated by a range of dangerous pollutants. Furthermore, pollution continues to cause massive fish kills around the country. Much of this pollution stems from virtually unregulated sources—polluted runoff from farms, mines, logging, and urban development.

Worse still, while we have made at least some progress in addressing chemical insults, we seem to be moving in the wrong direction for restoring the physical and biological integrity of the nation's waters, as Congress directed in 1972 as the overriding mission of the act. An alarming number of aquatic species are in serious jeopardy, and many have become extinct since the Clean Water Act was passed. Other populations are plummeting or face serious health effects due to the presence and accumulation of dangerous toxic pollutants in aquatic ecosystems. Major aquatic habitats, such as wetlands, floodplains, spawning and rearing grounds, and estuaries, continue to be paved over, drained, filled, dammed, channelized, and otherwise altered by the onslaught of human economic activities.

In summary, while much progress has been made, we are still a long way from meeting the most basic goals of the Clean Water Act. A large percentage of waters remain unsafe for fishing, swimming, and other uses. We continue to release toxic pollutants in toxic amounts and other pollutants in amounts that jeopardize aquatic species and whole ecosystems. And we are even further than before the act was passed from the mission of restoring the biological integrity of the nation's surface waters.

Chapter 3
The Economics of Clean Water

The previous two chapters identified some of the values at stake if the goals of the Clean Water Act are not met. Some are practical, such as our basic needs for safe water and ample seafood supplies and our desire to avoid getting sick from swimming in polluted waters. Others combine ethical and spiritual values, such as our moral responsibility to protect aquatic species from the ravages of human development, and the intangible benefits of sitting by a pristine lake or a free-flowing river amid the presence and diversity of the life that inhabits our aquatic treasures.

Even more than during the 1970s, however, when Congress and the administration debated whether the cost of clean water was a sound investment, this decade is one of limited financial as well as environmental resources. While the 1972 presidential candidates debated the price of war versus the price of domestic needs, the 1990s bring a time of relative military security in the face of a staggering federal deficit. Once again, the question will be raised: "Can we afford clean water?"

Because so much of the worth of clean water and healthy aquatic ecosystems transcends economic values, this question cannot be answered fully or definitively. No matter how hard economists try to value clean water, they will always miss much. This is not a criticism of economists or economics. It simply reflects that the unit of economic value, that is, money, does not match up fully with the units of environmental value, just as the number of canvases in the Louvre—or the market price of those works—do not convey the spiritual and intellectual value locked into those paintings. Indeed, there is no market at all for many of the values provided by aquatic resources.

Still, future debates over the Clean Water Act will continue to raise the issues of cost versus economic benefits and of need versus budgetary reality. It is not the intent of this chapter to suggest a formal cost-benefit analysis; even if we believed such a test were justified, it would be far beyond the scope of this report. It is essential, however, to address the economics of clean water as completely as possible, given the limitations of economic valuation. We can evaluate this issue from two opposite but related perspectives. From one angle, we can try to identify the economic value of our existing aquatic resources. From the other, we can try to get a sense of how much our economy has suffered from past destruction of these resources. Either way, past investments in clean water programs have reaped considerable benefits, and future investments are more than justified on economic as well as other grounds.

It is also useful to evaluate estimates of the future cost of clean water programs and to evaluate proposals for how to fund them. However much value we may place on our water resources, we must find a way to pay for their protection within the constraints of the budget deficits faced by local, state, and federal governments alike.

Economic Values of Water Resources

Not all of the values provided by clean water and healthy aquatic ecosystems can be *monetized,* that is, translated into dollars and cents. Some of the most important economic values of clean water, however, can be quantified to some degree. This exercise shows that water resources support tremendously valuable parts of the U.S. economy, such as commercial fishing and shellfishing, recreational fishing and hunting, other forms of water-based recreation and tourism, and water quality and flood control. As a result, millions of U.S. citizens enjoy jobs that rely on adequate protection of our waters.

Commercial Fisheries

More and more people are turning to fish and shellfish as a major source of protein. From 1972 to 1990, annual U.S. seafood consumption jumped from 12.5 to 15.5 pounds of fish per person.[1] The U.S. Department of Agriculture recently projected that by the turn of the century, this consumption rate could increase to as much as 17.2 pounds per person. If this forecast is correct, supplies of edible seafood products will need to increase by 18 to 31 percent by the year 2000 to meet the demand.[2]

A major and growing U.S. industry supports this demand. In recent years, the United States has ranked sixth among major fishing nations, behind the (former) USSR, China, Japan, Peru, and Chile.[3] The U.S. commercial fish harvest and the economic value of each year's catch generally increased between 1976 and 1990 (see Figure 3.1). Moreover, while the dockside value of U.S. landings in 1990 was $3.6 billion, even more economic value is generated through processing and wholesaling. That year, the industry's total contribution to the U.S. gross national product (GNP) exceeded $16.5 billion.[4]

Besides providing food, commercial fisheries also provide employment for hundreds of thousands of people per year. In 1988 alone, 363,703 people

FIGURE 3.1

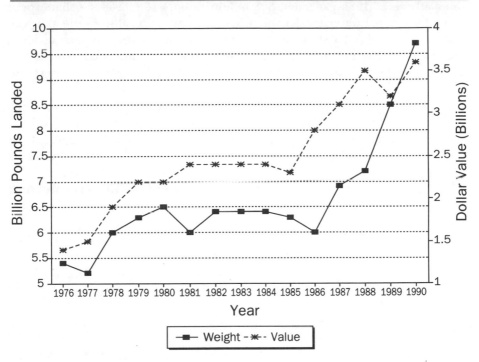

U.S. Fish Landings

Sources: NOAA, *Fisheries of the United States* (1990); Annual Fisheries Reports.

worked for commercial fisheries (fishing, processing, and wholesaling)—a major increase from 230,387 jobs in these industries in 1972. As fish consumption rises, so will the employment rate. Protecting clean water is essential to protecting these jobs.[5]

Recreational Fishing and Hunting

Fishing is the premier outdoor recreational activity in the United States. Over the past thirty years, the number of participants in fishing activities has increased steadily. In 1985, one in four people in this country sixteen years and older (46.4 million total) fished for pleasure. They took more than 870 million fishing trips and spent a total of 977 million days on the water.[6]

According to EPA, another 12 million youngsters (under sixteen years old) fished for pleasure in 1985.[7]

Recreational fishing, then, is a major U.S. industry. Total expenditures for this pastime in 1985 exceeded $28 billion dollars (freshwater fishers spent $19.4 billion; saltwater fishers spent $7.2 billion; $1.5 billion were unspecified).[8] Of this, $13.3 billion went toward trip-related expenses such as food, lodging, and transportation, while $13.6 billion went toward equipment and $1.3 billion toward other expenses such as fishing magazines, licenses, and membership dues.[9]

Shellfishing is also a popular water-based sport. In 1985, an estimated 4 million people (2 percent of the U.S. population) aged sixteen or older participated in recreational shellfishing, spending a total of 28 million days on the sport. Of the shellfishers, 50 percent lived in the South and accounted for 59 percent of all shellfishing days. Shellfishers estimated that they spent $2.3 billion to participate in the sport: $1.1 billion toward trip-related expenses such as food, lodging, and transportation; $1.2 billion toward equipment; and $63 million toward other expenses.[10]

Another common water-based sport is waterfowl hunting. According to the National Survey of Fishing, Hunting, and Wildlife Associated Recreation conducted by the Department of the Interior, 3.2 million hunters spent 25.9 million days hunting waterfowl in 1985. Expenditures for waterfowl hunting totaled more than $783 million (see Table 3.1).[11]

All forms of water-based hunting and fishing in the United States are on the rise. The total number of fishers and waterfowl hunters increased dramatically over the thirty-year period from 1955 to 1985, significantly raising expenditures to participate in these sports.

Other evidence confirms these trends, indicating that the economic value of clean water is on the rise:[12]

● The amount of fishing and boating on federal lands rose dramatically from 1982 to 1989.

● Total annual fishing licenses increased from 31 million to almost 37 million from 1970 to 1988; expenditures for those licenses rose accordingly.

● Sales of fishing tackle rose from $540 million in 1980 to almost $740 million in 1989.

By the turn of the century, about 84 million people in the United States are expected to participate in recreational fishing. And according to some estimates, the available living aquatic resources in the United States are limited and not expected to satisfy demands for quality recreational fishing by the year 2000.[13]

TABLE 3.1

Comparison of Major Findings in Fishing and Waterfowl Hunting		
	1970	**1985**
Total fishers	33.2 million	45.3 million
Freshwater	29.4 million	39.1 million
Saltwater	9.5 million	12.9 million
Waterfowl hunters	2.9 million	3.2 million
Fishing expenditures	$5.0 billion	$28.6 billion
Hunting expenditures	$244.5 million	$783.3 million
Days spent fishing	706.2 million	1.1 billion
Days spent hunting	25.1 million	25.9 million

Source: FWS, 1985 National Survey of Fishing, Hunting, and Wildlife Associated Recreation (1988), 150.

Other Water-Based Recreation

Measured by the number of trips away from home, our most popular out-door activities in 1987 included swimming outdoors (461 million trips per year), warm-water fishing (239 million trips per year), and motor boating (220 million trips per year).[14] People enjoy a broad array of activities near the waterfront. In 1985, 14.8 million people visited lakes or streams, 8.4 million people went to marshes or wetlands, and 5.7 million went to the ocean.[15] (According to EPA, the Interior Department reported an even higher number of outdoor [nonpool] swimmers in 1982.)[16] While at these sites, the majority of people enjoyed feeding, photographing, and observing birds and mammals. Others enjoyed observing and feeding fish, amphibians, reptiles, shellfish, and marine mammals. Other recreational activities included water-skiing, walking along the shore, kayaking, canoeing, rafting, surfing, sunbathing, and picnicking.

Forms of water-based recreation other than hunting and fishing are on the rise as well. For example:

● Visitors to national seashores and lakeshores jumped from 18 million in 1981 to more than 23 million in 1988.

● The number and value of recreational boats in the United States almost doubled from 1970 to 1989, with total expenditures on recreational boating quadrupling.[17]

Other water-based recreation, like hunting and fishing, generates substantial and increasing economic benefits. Direct and indirect expenditures on boating (boats, motors, equipment, fuel, insurance, and so on) doubled from $11 billion in 1970 (1990 dollars) to almost $20 billion in 1988 (although expenditures dropped to just under $14 billion—still a 25 percent increase over 1970—in recessionary 1989–90). This activity supports more than sixty-two hundred manufacturers of boats, trailers, motors, and accessories, as well as more than eighty-three hundred marinas, boat yards, and yacht clubs. The recreational marine industry as a whole provides jobs for about six hundred thousand people.[18]

While similar nationwide economic data are not available for swimming expenditures, a case study in Florida highlights the tremendous importance of tourism related to this sport. According to this study, Florida beach users generated a total of $2.3 billion in economic benefits in 1984. This included direct expenditures of $1.8 billion (almost 1.5 percent of total gross sales in the state), more than $400 million in payroll for about eighty-four thousand jobs, and tax revenues of almost $95 million.[19]

Intangible Values

Measuring actual expenditures is only one way to establish the value of clean water to people. Aesthetics, nature, and the opportunity to view wildlife are other important values of pristine waters. As the previously cited statistics indicate, from swimming to boating to sunbathing, being near the water is a favorite way for people to spend their leisure.

Many of these activities contribute to the economy directly through equipment purchase, travel expenditures, and activity fees. It is difficult, however, to place a value on the pleasure of spending a sweltering summer day in or near a cool, clean body of water, or on the satisfaction of seeing wildlife in their natural habitat. Placing a price tag on the assurance that these resources will be left for our children as well is even more difficult.

Some economists are beginning to assess economic values to intangible environmental benefits through techniques such as contingent valuation surveys. These surveys generally describe a hypothetical market in which a public good may be purchased and ask participants how much they would be willing to pay for an increase in the level of this public good.

Care must be taken interpreting the results of contingent valuation studies. Uncertainties about this relatively new method lead to questions about how complete or accurate such valuation can be. For example, if survey respondents say they would pay less for ecological or aesthetic resources during a recession than during a period of economic growth, does this mean the resource is really worth less? Moreover, a respondent's willingness to pay for a resource is likely colored by his or her knowledge and understanding about the benefits provided by that resource. A respondent who does not know that wetlands support many endangered species, for example, probably would be willing to pay less to protect an acre of wetland. Economists continue to debate contingent valuation methodology, including whether respondents should be surveyed based on their current knowledge or whether they should be given more information.[20]

Despite its limitations, economists increasingly are turning to contingent valuation to place some monetary value on environmental benefits that otherwise would escape evaluation. These studies can at least fill in some of the gaps left by traditional economic methods in valuing our water resources.

In a 1991 contingent valuation study, for example, Resources for the Future asked a sample of U.S. residents to value a set of water quality improvements. Respondents were to give a monetary value to minimum levels of boatable, fishable, and swimmable water. Those who gave usable answers were willing to pay an average of $106 per year for boatable water quality, plus $80 for fishable minimum water quality and an additional $89 to reach a national minimum of swimmable water quality. On the average, people were willing to pay a total of $275 per year to ensure clean water.[21] Based on these answers, the economic benefit of achieving the national swimmable water quality goal was estimated at $29.2 billion per year in 1990. However, a range of $24 billion to $43 billion per year was considered reasonable. When the results of this study and similar surveys were aggregated and adjusted to correct for the number of current households and the consumer price index, the estimated annual value of clean water became $46.7 billion.[22] Clearly, the public places a large economic value on even the intangible benefits of our aquatic resources.

To some extent, increased values of waterside property confirm public willingness to pay for the intangible values of water resources. The Environmental Protection Agency cites the following examples:

- A 1983 study estimated that properties on polluted St. Albans Bay on Lake Champlain, Vermont, were selling for $4,500 less than similar properties elsewhere on the lake, with a total property value loss of $2 million.
- Water quality improvements in the lower Willamette River near Portland, Oregon, led to residential property increases of 16 to 25 percent.
- Reduction of water pollution in San Diego Bay, California, in the 1960s resulted in an approximately 8 percent rise in residential property values.[23]

Wetlands

Because their importance has been hotly debated in recent years, it is particularly important to underscore the economic, as well as the environmental, values of wetlands. While some may view wetlands as just swamps, or even as eyesores, they are actually rich areas that support an abundance of wildlife and provide services that may be worth billions of dollars in economic benefits. For example, an estimated 50 million people spend nearly $10 billion annually to observe and photograph wetland-dependent birds.

Wetlands also sustain much of the country's seafood. In the Southeast, 96 percent of the commercial catch and more than 50 percent of the recreational catch is dependent on estuarine and other coastal wetlands. Estimates of the value of coastal wetlands to commercial and recreational fisheries range from about $2,200 per acre on the Pacific coast to almost $10,000 per acre along parts of the Florida coastline.[24] Wetlands also provide habitats for valued mammals, such as muskrat, beaver, and mink. Muskrat pelts alone are worth more than $70 million annually.[25]

A 1981 case study by researchers at Tufts University, Massachusetts, illustrates the economic importance of wetlands.[26] The researchers evaluated 8,535 acres of wetlands in the Charles River basin, Suffolk, Norfolk, and Middlesex counties, Massachusetts. The economic benefits measured were flood control, increases in land value, pollution reduction, water supply, and recreation and aesthetics. Other values, including preservation and research, vicarious consumption and option demand, and undiscovered benefits, were "described, not monetarized."

In a 1976 study of flood control in the Charles River watershed, the Corps had estimated that wetlands provide 75 percent of the basin's natural water storage. Accordingly, the loss of all of the wetlands used in the Tufts case study would produce expected annual flood damage of almost $18 million, or about $2,000 per acre (1978 dollars).

Wetlands also provide a wealth of recreational opportunity. The Tufts case study used two methods to estimate the monetary value of recreational activities such as small-game hunting, waterfowl hunting, trout fishing, warm-water fishing, and nature study. The first considered factors such as the days of use per acre per year, the actual expenditures per user per year, and willingness to pay per user per year. According to this method, the recreational use of one acre of wetland in the study areas was estimated to be $187.74 per year.

The second analysis of the recreational value of these wetlands estimated the price at which people would sell their rights to recreate in wetlands (a variation on contingent valuation, described previously). This method estimated a total annual recreational value of $3,366 per acre of wetland. The present

FIGURE 3.2

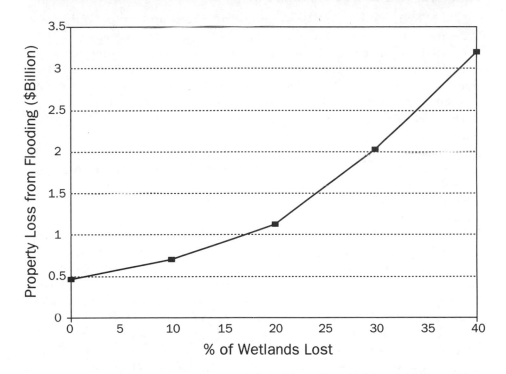

Annual Property Losses Due to Wetlands Water Storage Loss
(For 8,535 Wetlands Acres in the Charles River basin, Massachusetts)

value of this recreational worth (in 1978 dollars), generated by adding this annual income stream infinitely into the future (discounted at 6 percent), was $56,100 per acre.

Although the study produced conservative estimates of the monetary value of wetlands, its findings are revealing. Following is a breakdown of the estimated economic benefits of 1 acre of Charles River basin wetland in 1978 dollars: flood prevention, $33,370; local amenity, $150–$480; nutrient reduction, $16,960; water supply, $100,730; recreation, $2,145–$38,469. The total long-term worth of just 1 acre was estimated to lie between $153,000 and $190,009—not bad for "just a swamp."

Case Study: The Chesapeake Bay

In 1987, the Maryland Department of Economic Employment and Development conservatively estimated that the Chesapeake Bay was worth $678 billion to the economies of Maryland and Virginia.[27] This figure was the sum of two parts. First was the total value of the annual activities that could not take place without the Bay. Second was the value of premiums that people were willing to pay for waterfront, water-view, or water-access homes.

The study estimated that the 1987 impact from commercial fishing, port activities, ship and boat building, ship repair, and Bay-related tourism was $31.6 billion. The 1987 present value of this income stream, discounted into the future at a rate of 5 percent, was $632 billion. The annual value of commercial fishing in the Chesapeake Bay region was $520 million (1987 dollars), subject to annual fluctuations. This included the actual dockside value of the fish plus the processing value generated by businesses that prepare fresh or frozen fish and other seafood for shipment. Tourism accounted for an additional $8,396 million, including recreational activities such as boating, fishing, hunting, sight-seeing, and dining on regional cuisine. Port activities accounted for $5,344 million per year, and ship building and repair yielded $17,346 million per year. The total land premium for waterfront, water-view, or water-access land around the Bay was estimated at $46.2 billion. Premiums for Bay area properties are different depending on the neighborhood and size of the lot and can be as high as $250,000 per lot in prime locations. Even in more remote locations, realtors assess premiums of $35,000–$50,000 for waterfront, water-view, and water-access lots. This study used an average premium of $50,000 per site.

Economic Losses Due to Water Resource Degradation

A different way to view the value of clean water is to look at the economic loss that occurs when water resources are degraded. (Since some of the values discussed in this section address the loss of positive values previously addressed, we do not imply that estimated monetary values should be added together with estimated losses.) Loss of natural and economic resources occurs in a number of ways: Pollution causes sources of drinking water and seafood supplies to be closed or restricted or increases treatment costs for drinking water; water and seafood contamination causes human illnesses, with attendant medical costs or lost productivity of the work force, as well as human suffering; flooding and erosion devalue property; and habitat destruction and chemical pollution decrease valuable recreational opportunities. Less obvious, but equally important from an economic perspective, is the loss of valuable renewable resources, such as reusable water and sewage sludge, when toxic pollutants contaminate them. As with clean water and aquatic resources, too little work has been done to quantify the value of these economic losses. Some information is available, however, to underscore the tremendous current and potential future economic losses we face if we do not protect our aquatic resources.

Losses Due to Pollution

The most obvious and immediate economic losses due to water pollution and lost or degraded aquatic habitat occur when commercial fisheries decline or are closed due to contamination. As discussed in chapter 2, major commercial fisheries have already been lost or are declining severely. While little information is available to quantify these losses directly, regional studies exemplify their severity.

One study, for example, evaluated the impact of pollution on the economic value of shellfish resources in the states of New York, Rhode Island, Massachusetts, and Connecticut. More than 75 percent of Connecticut's bays, harbors, and estuaries are restricted to shellfishing; 25,000 acres in southern Massachusetts are similarly restricted; and the harbor of New Bedford, Massachusetts, has been declared a Superfund site (listed federally as being of high priority for cleanup) because of pollution by PCBs. In 1990, the dockside value of commercial landings in these states exceeded $72 million. Using a conservative economic multiplier of 4.5, indirect revenues could approximate $325 million. The estimated annual value assigned to the potential harvest of shellfish from prohibited areas (areas closed due to pollution or other unsafe conditions) is approximately $36 million. If these shellfish resources could be utilized, the value of commercial landings would increase by about 50 percent, and total indirect revenues might be as high as $487 million.[28]

Other studies cited by EPA present a similar picture:[29]

● A ban on commercial fishing of striped bass in New York Harbor due to PCB contamination reportedly costs the area about $15 million a year.

● Eliminating pollution in the Delaware River estuary could bring back a commercial shad fishery worth up to $431,000 per year.

While no national data are available to quantify the monetary value of pollution-caused commercial fishery losses, clearly these economic impacts are substantial.

Losses Due to Flooding and Erosion

Proper investment in and protection of aquatic resources could also prevent the burden of flood management and erosion control. Flooding is the most costly natural hazard in the United States, causing a greater loss of life and more property damage than all other natural disasters combined. In 1991, the Federal Emergency Management Agency (FEMA) estimated that there were about 94 million acres of flood-prone land in the United States.[30]

Between 1980 and 1985, flooding caused more than $3 billion in damages and the loss of an average of 191 lives (which cannot be measured by economic value) per year.[31] These statistics were the same or higher for each five-year period since 1970. Moreover, they greatly exceed those for the fifty years preceding 1970. Loss of wetlands may be in part responsible for the rise in damages and deaths due to flooding. "Studies in Wisconsin, for example, showed that watersheds consisting of 30 percent or more of wetland had 60 to 80 percent lower floodpeaks than watersheds with little or no wetland."[32]

Like flooding, erosion is a serious condition that affects the country's shorelines. The Corps determined that erosion is a problem for 574,000 miles of streambanks, a serious problem for 142,100 miles of them. About 78 percent of all streambank erosion takes place west of the main stem of the Mississippi River. In addition to destroying natural habitat, streambank erosion causes millions of dollars in economic damages annually (see Table 3.2).

Erosion is especially serious for oceanfront property. Beach erosion in the New York to Washington corridor averages 2 to 3 feet per year. In 1986, North Carolina found that 750 oceanfront structures insured at approximately $50 million would be lost to erosion in the next ten years; the potential exists for 5,000 structures to be lost over the next sixty years, with 4,200 structures at immediate risk in the event of a major coastal storm.[34]

Losses of Drinking Water

Eliminating drinking water contaminants is important to protecting public health. It can also reduce overall costs of supplying safe drinking water—costs borne by public water supply users rather than by those who pollute the water in the first place (although in some cases these groups overlap).

Contaminants in public drinking water supplies increase treatment costs. Some costs to ensure a safe drinking water supply are inevitable, especially

TABLE 3.2

Streambank Erosion (Average Annual Damages in Millions of $)			
Arkansas-White-Red	93.5	Ohio Valley	5.7
California	56.2	Colorado Basin	4.8
Lower Mississippi	38.9	Great Lakes	2.7
Pacific Northwest	23.6	Souris-Red-Rainy	1.2
Missouri Basin	16.8	Alaska	1.2
South Atlantic Gulf	11.8	New England	1.8
Middle Atlantic	10.9	Tennessee Valley	0.9
Rio Grande	10.5	Great Basin	0.5
Texas Gulf	7.8	Hawaii	0.0
Upper Mississippi	5.8		
TOTAL			294.0[33]

Source: Federal Interagency Floodplain Management Task Force, *Floodplain Management in the U.S.* (1992), 18–19.

since drinking water contaminants occur even in relatively natural water supplies. But the presence of pesticides, other artificial organic chemicals, and other human-introduced pollutants results in the need for expensive new treatment technologies at drinking water supplies around the country. No national study quantifies these costs completely or evaluates how much money could be saved by protecting drinking water supplies in the first place. But available information suggests that preventing pollution of drinking water supplies is a sound economic as well as public health investment.

One EPA study, for example, the National Survey of Pesticides in Drinking Water, indicates that about 50 million people in the United States live in areas most vulnerable to pesticide contamination of groundwater. According to this study, protecting the water supply can avoid estimated remedial costs of $250 to $750 per household per year. These costs may total as much as $100 million to $150 million per year. In addition, people who live in areas subject to drinking water contamination are willing to pay as much as $50 per household per year for programs that evaluate groundwater problems and identify efforts to resolve them. The total amount generated from this type of program could equal as much as $1 billion per year from households at risk from pesticide contamination.[35]

Similar results are apparent for drinking water systems in major cities, where supplies often come from surface waters. The estimated costs of filtering the New York City water supply for traditional pollutants (pathogens and solids) range from $1.5 billion to $5 billion for construction, with operating costs of $300 million per year. If activated carbon is needed to filter toxic contaminants, however, construction costs double, up to $10 billion, and operating costs triple, up to $900 million per year.[36]

Some data are available to characterize drinking water treatment costs on a national scale. For example, the General Accounting Office estimated annual costs for implementing amendments to the Safe Drinking Water Act that require stricter monitoring and reporting of drinking water treatment (see Table 3.3).[37]

Losses of Recreational Opportunities

In chapter 2 we documented the widespread loss of recreational opportunities resulting from beach closures, fishing bans and advisories, and similar impacts. No comprehensive data are available to quantify the economic value of these losses. The information that is available, however, suggests that these losses are large, and can have devastating impacts on local economies.

According to EPA, for example, charter fishers in Lake Michigan reported that sales dropped by up to 40 percent after issuance of a report citing the health risks from eating contaminated Great Lakes sport fish. Similarly, EPA cites a 1990 *Los Angeles Times* report that about one hundred thousand Californians abandon sportfishing each year due to population declines caused by habitat loss and because of health risks from polluted fish.[38] If these impacts are multiplied across the thousands of fishing bans and advisories in place around the country, the economic impacts could be quite severe.

Similar effects are evident from the thousands of beach closures and warnings that have plagued U.S. coastlines in recent years. A study by the New Jersey Bureau of Economic Research, Rutgers University, and the New Jersey Department of Environmental Protection found that beach closures in New York and New Jersey in 1988 cost an estimated $2 billion in lost tourism dollars.[39]

TABLE 3.3

Estimated Annual Costs to Water Systems for Implementing 1986 Amendments to the Safe Drinking Water Act (Millions of 1986 $)

STANDARD	NUMBER OF SYSTEMS AFFECTED	ANNUALIZED CAPITAL/ O&M COST	AVG ANNUAL MONITORING COST	TOTAL COMPLIANCE COST
Volatile organic chemicals	1,824	$ 32.7	$ 23.1	$ 55.8
Filtration	10,228	511.6	17.1	528.6
Total coliforms	200,183	0.0	75.2	75.2
Synthetic organic chemicals	2,284	45.4	33.2	77.5
Inorganic chemicals	1,896	123.2	12.4	135.6
Lead/copper corrosion control	43,927	302.2	32.9	335.2
Radionuclides	22,867	790.3	2.6	792.9
Disinfection	103,354	474.8	12.8	487.7
TOTAL		$2,280.2	$208.3	$2,488.5

Source: GAO, *Drinking Water: Compliance Problems Undermines EPA Program as New Challenges Emerge*, GAO/RCED-90-127 (1990), 53.

Losses of Potentially Reusable Resources

The contamination of domestic sewage by toxic pollutants from industrial, household, and other sources is part of a cascade of problems that causes pollution of air, land, and water. By contaminating these potentially valuable materials, we lose economically valuable renewable resources that could be used to fertilize and irrigate crops, forests, and other lands. The pattern looks like this:

1. Our failure to reduce or eliminate the release of toxic chemicals into our public sewers causes contamination of air (through evaporation of volatile organic chemicals), sewage effluent, and sewage sludge.

2. We discharge sewage effluent into our rivers, lakes, and coastal waters, causing serious water pollution problems.
3. We treat sewage sludge as a waste to be burned in incinerators (causing more air pollution) or to be dumped in landfills (polluting ground or surface water).
4. Instead of using the valuable water and soil nutrients in sewage effluent and sludge, we drill for oil, manufacture chemical fertilizers, and build massive new dams and other water projects, all of which cause even more environmental harm.

Steps to eliminate contamination of our sewage systems thus could reduce environmental impacts while freeing up valuable materials that would lower water supply and fertilizer costs around the country. While it is difficult to assess a precise national value on these losses, information suggests that we are currently losing a tremendous economic resource by polluting rather than reusing these materials.

Wastewater

In 1985, approximately 31 billion gallons of sewage wastewater were discharged to surface water every day.[40] Put in perspective, this amounts to more than 11 trillion gallons a year, or more than one-third of total annual consumptive U.S. water use. Much of this wastewater could be reused productively if it were reclaimed properly and if it were sufficiently free of toxic contaminants. Properly reclaimed wastewater can be used for groundwater recharge, industrial processes, irrigation, recreational lakes, or direct municipal reuse.

University of California researchers suggested five important economic and environmental reasons to use properly reclaimed wastewater. First, we can reduce or eliminate the costs of building additional water supply, collection, storage, and transport systems. Second, we can reduce or eliminate the serious environmental harm from building these additional systems. Third, the reuse of wastewater helps to conserve scarce water resources. Fourth, wastewater reuse can reduce the rate of groundwater depletion for areas that obtain their water from wells. Fifth, we can protect water resources from pollution by reducing total sewage discharges.[41]

An additional reason not cited in this study is that wastewater discharges are rich in nitrogen, phosphorus, and other nutrients needed to grow various crops. Replacing these nutrients uses scarce oil resources that we convert at petrochemical factories into chemical fertilizers.[42] Moreover, sewage reuse can be better than chemical fertilizers for agriculture because reclaimed sewage releases nutrients more slowly and continuously, not in a single large dose.

Nor would reclaimed sewage run off as readily into local surface water and groundwater as chemical fertilizers do. In test plots in India and Thailand, rice, wheat, and cotton yields were considerably higher when the crops were grown with reused wastewater than in similar plots using well water and commercial fertilizer.[43]

We do use some wastewater in this country. In more than a thousand wastewater reuse projects in the United States, water is reused for irrigation, industrial cooling and processing, and groundwater recharge. By the late 1980s, Los Angeles and Orange counties, both in California, reused about 70 million gallons of water per day—about 15 percent of the area's wastewater volume—for irrigation, cooling water, and groundwater recharge. But despite these efforts, reclaimed water meets only about 0.2 percent of water use needs in this country. By comparison, reclaimed water met 4 percent of Israel's water needs in 1980 and is expected to reach 16 percent in that country by the year 2000.

To make better use of reclaimed wastewater, we need to answer two critical questions: First, what reclamation (treatment) levels are necessary to protect human health and the environment? Second, is wastewater reuse cost-effective relative to new water supplies or water conservation? That is, what are the economic benefits of wastewater reuse?

To answer the first question, note that different uses of water carry different potential for human exposure—from uses such as industrial cooling water, which require little or no human contact, to direct human consumption. Depending on the category of reuse, wastewater may be treated at the primary, secondary, tertiary, or advanced level (see Table 3.4). Primary treatment is the removal of suspended solids to designated levels; secondary is the reduction, by activated sludge process, of the biological and chemical oxygen demand of wastewater to regulation levels; tertiary is the specific treatment process of coagulation, clarification, and filtration to improve the effectiveness of disinfection; and advanced includes reverse osmosis to lower the dissolved solids and activated carbon to absorb organic compounds.

Clearly wastewater reuse is controversial. "There is legitimate concern for fail-safe disinfection technology when reclaimed water is to be used for purposes involving close human contact where inhalation or ingestion are possible. This is of considerable concern because the majority of waterborne diseases reported can be traced to water treatment plant failure."[44]

But uses of reclaimed wastewater that do not involve direct human ingestion over a long period of time seem to pose few if any direct health risks. Public opinion surveys confirm this view. Seven studies performed in 1988 tested the percentage of opposition for different uses of reclaimed water.[45] The majority of respondents opposed only the uses where contact is very likely (see Table

TABLE 3.4

Categories of Wastewater Reuse and Associated Treatment Requirements

REUSE CATEGORY	TREATMENT
Groundwater	
Spreading	Tertiary
Injection	Advanced
Industrial	
Human contact unlikely	Secondary
Human contact likely	Tertiary
Irrigation	
Fodder and fiber	Primary
Food crops	Secondary
Parks and playgrounds	Tertiary
Recreational lakes	
Restricted	Secondary
Nonrestricted	Tertiary
Direct municipal	
Potable	Advanced & complete treatment
Lawn irrigation	Tertiary

Source: William H. Bruvold, Betty H. Olson, and Martin Rigby, "Public Policy for the Use of Reclaimed Water," *Environmental Management* 5, no. 2 (1981), 97.

3.5). Thus, reclaimed water should start to be used for those purposes where risks are low and public opinion is high.

The second key question is whether wastewater reclamation and reuse is cost-effective, that is, whether the added treatment and distribution costs necessary to use reclaimed water exceed the cost of other water sources (new supplies or water efficiency). Some studies have shown that wastewater reclamation and reuse can provide significant economic benefits.[46]

TABLE 3.5

Weighted Mean Percentage Opposed to Twenty-seven Uses of Reclaimed Water

TYPE OF REUSE	DEGREE OF CONTACT	% OPPOSED
Restaurant food preparation	Very High	56
Drinking water	Very High	54
Cooking in the home	Very High	48
Preparation of canned vegetables	Very High	46
Bathing in the home	High	33
Pumping down special wells	High	27
Home laundry	High	23
Swimming	High	21
Commercial laundry	High	19
Spreading on sandy areas	High	16
Irrigation of dairy pasture	Moderate	14
Irrigation of vegetable crops	Moderate	13
Vineyard irrigation	Moderate	13
Orchard irrigation	Moderate	10
Pleasure boating	Low	10
Hay or alfalfa irrigation	Low	8
Commercial air conditioning	Low	
Golf course lakes	Low	6
Electronic plant process water	Low	5
Home toilet flushing	Low	4
Lawn irrigation	Low	4
Park irrigation	Low	3
Golf course irrigation	Low	3
Road construction	Low	2
Stream or river discharge	Low	—
Bay or ocean discharge	Low	—

Source: Bruvold, Olson and Rigby, "Public Policy for the Use of Reclaimed Water," *Environmental Management* 5, no. 2 (1981), 95.

Distribution and treatment costs, including energy costs, are unique to each situation. For example, the cost estimate for spreading groundwater is low because the process involves little equipment. On the other hand, groundwater injection involves using pumps and additional equipment, which brings its distribution cost to the moderate level. Any evaluation of the economics of a particular treatment and reuse option must take into consideration a wide range of variables.

Because of these variables, it is difficult to place a national price tag on the value of reclaimed water. But consider this: Even if savings were only a penny a gallon over alternative sources of water and fertilizer, the 11 trillion gallons of wastewater that we dump into our surface waters each year would have an astonishing potential economic value of $110 billion. This does not even account for the economic and environmental harm caused by existing sewage pollution. Clearly, society could save on both the economic and environmental costs of providing water and fertilizers and on the ongoing environmental costs of water pollution from sewage releases into our surface waters by investing in better pollution controls so that we can reuse rather than discharge sewage wastewater.

Sludge

Sewage sludge, or *biosolids,* is another sewage by-product that is not being used to its full potential, in part due to contamination by toxic pollutants. As with wastewater, contamination of sewage sludge leads to a related pattern of economic and environmental problems: We burn sludge in incinerators or dump it in landfills, causing air or water pollution, and we throw away valuable nutrients and soil conditioners, which we must then replace with artificial fertilizers made from limited petroleum resources.

Sewage sludge is the solid, semisolid, and liquid residue removed from wastewater during treatment. It is made up mostly of water, organic matter, and nutrients, such as nitrogen, phosphorus, zinc, calcium, and magnesium, with some trace elements, which, in small quantities, benefit plant growth. According to an EPA estimate, sludge typically contains $50 to $60 worth of nutrient and soil conditioning value per dry ton.[47] Without adequate pretreatment and sewage treatment, however, it may also contain bacteria and toxic and other pollutants. Heavy metals and other contaminants can pose human health and environmental risks if present in sufficiently high quantities, underscoring the need for adequate sludge quality standards and pretreatment requirements.

Because the nutrients in biosolids are a valuable economic resource, thousands of municipalities are recycling sludge for agricultural and other uses. It has been used successfully in the production of food, feed, horticultural crops, and sod; turf maintenance; reclamation of mined lands; and forest production. For example, in 1988, all sludge from Washington, D.C., was used to condition or fertilize land. More than one-third was used on 7,000 agricultural acres in Virginia, and some 30,000 tons were applied in Maryland.[48]

However, on a national basis, less than 40 percent of all sewage sludge is used beneficially. The rest is discharged into water bodies, incinerated, or place in landfills.

Sludge nutrients benefit agriculture in the form of fertilizer. A 1985 study showed that cotton lint yields were comparable to those obtained with commercial fertilizer when sludge was used at rates of 20 to 80 milligrams per hectare (mg/ha). At the high application rate of 466 mg/ha of solids, grain and fodder yields were the same as with commercial fertilizer. Sludge application on corn also yielded similar results as with commercial fertilizer. Moreover, snap bean yields increased three-fold when grown using sludge rather than commercial fertilizer.[49]

A 1991 study on sludge conducted at the University of Florida Institute of Food and Agricultural Sciences demonstrated the successful use of municipal wastes, including sewage sludge, to grow vegetables and tropical crops.[50] Processed Municipal Waste (PMW) includes processed sludge, yard trash, newsprint, and food wastes. This study evaluated three different PMWs: Daorganite (processed from sewage sludge by Metro-Dade County); compost (sewage sludge mixed with yard trash and garbage); and Agrisoil (composted solid waste). The results were as follows:

Tomatoes. Yields of large, marketable, and total fruits were higher than in control plots with no PMW when sludge was incorporated into the soil or placed on plant beds at a rate of either 3 tons per acre (t/a) or 6 t/a. After flowering, tomato plants with 5 and 10 t/a of Daorganite had enhanced growth compared to untreated plants, especially in flooded areas of the field.

Squash. Low rates of processed municipal wastes resulted in higher squash yields.

Areca Palms. Plants were potted with 0, 15, 30, and 45 percent sludge or compost. After nearly a year, plants showed elevated height, width, and circumference in 15 percent Daorganite sludge; in 15, 30, and 45 percent compost; and in a combination of 15 percent Daorganite sludge and 30 percent compost.

Papayas. Sludge and compost applications increased the number of medium-sized fruit, but the total number of fruit per plant remained constant.

Sludge applied at 1.5 t/a and compost applied at 1.5 to 3 t/a increased medium and total fruit numbers compared to untreated plants.

Sludge also has a beneficial effect on soil.[51] Studies have shown that sludge and sludge-based compost decrease the bulk density of soils, rendering a better environment for plant root growth. Sludge also increases soil aggregation, which reduces the potential for erosion. One study showed that application of sludge reduced rates of runoff and sedimentation.

Water retention and hydraulic conductivity increase with sludge application. This allows plants better access to water, especially during dry periods or stress.

Sludge is also being used increasingly in forestry to revegetate and stabilize forests that have been damaged by harvesting, fires, landslides, and so on. Sludge helps to shorten wood production cycles by speeding up tree growth. According to studies at the University of Washington, using sludge as a forest fertilizer increases height and diameter from two- to ten-fold in various tree species.[52] Furthermore, trees grow twice as fast on soil that has been fertilized with sludge.

Sludge can also be used cost-effectively to revegetate areas destroyed by mining, dredging, and construction, to fertilize highway medians, and to cover landfills. In a strip-mined area in Illinois, reclamation using sludge cost $3,660 an acre, compared to $3,395 to $6,290 per acre for other methods.[53]

As with water, the total national economic value of sewage sludge is difficult to ascertain because of the many variables in sludge quality, transportation costs, and so on. But based on current prices of sewage sludge products, the market value of sludge currently being burned or landfilled is substantial. Heat-dried, high-nitrogen-content sludge with a consistent nutrient value is being sold in bulk for $90 to $190 per ton. More odorous products with less consistent nutrient content are sold in bulk for $20 to $60 per ton. Lower-nitrogen sludges sell in bulk for about $4 to $16 per ton or, after alkaline stabilization, for $2 to $10 per ton. In smaller quantities, sludge products retail for around $5 to $16 per 40-pound bag, equal to $250 to $800 per ton.[54]

Clearly, additional investment in clean water programs designed to reduce levels of toxics in sewage sludge, making them usable for fertilizer and soil conditioner for a range of applications, could yield substantial environmental as well as economic benefits.

Paying for Water Resource Protection

If water resources need and deserve greater protection, as suggested in chapter 2 and earlier sections of this chapter, why have larger investments in clean water programs not been made? The answer is complex and involves politics, competing social and economic needs, and severe budgetary constraints facing federal, state, and local governments. Full solutions to the problems of financing clean water programs are beyond the scope of this book. Nevertheless, it is important to take stock of the major funding needed to implement the Clean Water Act fully.

Shortfalls in Clean Water Funding

As shown in chapter 2, billions of dollars have been spent by federal, state, and local governments, as well as by private businesses, to implement the Clean Water Act over the past two decades. Total costs of water quality control, according to EPA estimates, have increased from about $9 billion in 1972 to almost $39 billion in 1990 (in constant 1986 dollars) and are expected to increase to more than $57 billion by the turn of the century.[55] Nevertheless, remaining funding needs are large. While no single source has analyzed these shortfalls comprehensively, some examples of serious funding problems have been identified:

- Environmental Protection Agency funding has not grown as fast as the proliferation of congressional mandates. The agency's operating budget was slashed badly in the early 1980s. While some amends were made in the late 1980s and early 1990s, EPA's overall operating budget grew by only 25 percent (in constant dollars) from 1981 to 1992, a time during which major new environmental laws were passed or in early phases of implementation (such as Superfund and the Resource Conservation and Recovery Act [RCRA]) and others were expanded dramatically (including the Clean Air and Clean Water acts).[56]

- State and tribal water quality programs also are underfunded. States have identified a $400 million shortfall in the funding they believe is necessary to manage state water quality programs. Native Americans believe that they need an additional $40 million for water resource programs nationally.[57]

- The nation's sewage treatment construction and repair needs continue to be staggering. The Environmental Protection Agency's most recent Needs Survey identified more than $110 million in sewage treatment needs over the next twenty years, counting only those projects that are eligible for federal funding. The states identify $138 billion in documented needs.[58]

- The EPA Needs Survey does not account fully for funding needed to address pollution from combined sewer systems in older U.S. cities that overflow raw sewage (CSOs, or combined sewer overflows) and other wastes when rainfall overloads them. Estimates of the cost of addressing this problem are as high as an additional $80 billion.[59]

- Some of the highest sewage treatment bills are concentrated in a few large cities, such as New York (estimated $10 billion in needs); Los Angeles ($5 billion); Cincinnati, Ohio ($2.5 billion); and Sacramento, Boston, San Diego, and Seattle (each above $1 billion). Rate hikes in these cities could have serious impacts on poor populations.[60]
- Other severe funding problems occur in small, rural communities where lack of access to bond markets and diseconomies of scale make modern sewer systems difficult to afford.[61]
- Important new clean water programs, such as Congress's 1987 requirement for cities to develop comprehensive programs to control polluted urban runoff, impose additional costs on many communities. Cities estimated that they spent $130 million to $140 million nationally to prepare their initial stormwater permit applications and that actual compliance costs will be much higher.[62]

Polluted Runoff Controls

Given that runoff pollution causes at least half of our serious water pollution problems, the most serious funding shortfall in all clean water programs is the lack of resources to tackle this serious problem. (We will discuss the problems caused by polluted runoff in more detail in the next chapter.) Congress has never fully funded the new nonpoint source pollution program under section 319 of the Clean Water Act; only about half of the funding authorized by section 319 was actually appropriated.

While industry and governments have spent billions to reduce pollution from factories, sewage plants, and other point sources, they have spent only a tiny fraction of these amounts to control polluted runoff. Moreover, EPA estimates show that we are going in the wrong direction. Each year, a smaller percentage of our total clean water funding has gone to controlling polluted runoff, and EPA predicts that this trend will continue over the next seven years (see Figure 3.3).

The Case for Increased Federal Funding

Given the pressing federal budget deficit, 1993 is not a good time to ask Uncle Sam for money. Some also argue that local individuals and businesses that cause water pollution should bear the costs of controlling it; where federal funds are not available, localities have the legal and moral obligation to pay for reducing or eliminating their own pollution.

On the other hand, state and local governments face funding constraints as severe as those faced by the federal government. Moreover, while in theory

FIGURE 3.3

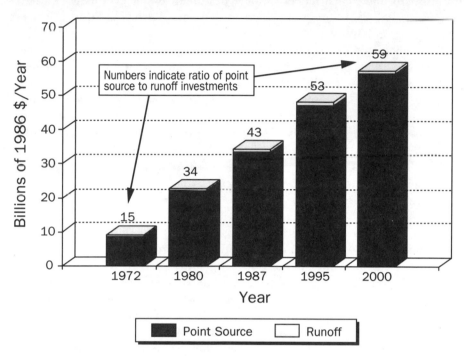

**Money Spent on Water Pollution Control
(Millions of 1986 $)**

Numbers indicate ratio of point source to runoff investments

Source: EPA, *Environmental Investments: The Cost of a Clean Environment*, at 3-3.
Note: Figures are for all programs, nationwide.

people should pay to control their own pollution, some are better able to afford increased rates than others. And, after all, the public as a whole shares the benefits of clean water and healthy aquatic ecosystems through improved human health, enhanced fish and wildlife populations, and better recreational opportunities. From the perspective of justice, then, there is also a good argument to be made for sharing at least some of the costs of nationwide water pollution control through federal funding.

From 1972 to 1987, through Title II of the Clean Water Act, Congress provided an average of $5 billion a year in construction grants to municipalities to build U.S. wastewater treatment infrastructure. Based in part on then-President Ronald Reagan's opposition to these funding levels, however, and in an effort to make each federal sewage treatment dollar go further, Congress shifted in 1987 from the grant program to a State Revolving Fund (SRF) program, to be capitalized at $8 billion.

The policy of shifting from grants to revolving loans is sound. There are a number of serious problems, however, with the SRF program as written and as implemented:

1. Congress has not yet appropriated all of the $8 billion promised to capitalize the state funds.
2. As discussed previously, this funding level is inadequate to meet the $200 billion in estimated wastewater needs over the next twenty years.
3. Small, rural communities have serious difficulties constructing needed sewage treatment facilities with low-interest loans alone.

Meeting these needs will be expensive. But since most people have difficulty imagining what a billion dollars means, it is useful to put them in perspective. While investments in environmental improvement generally may seem large, EPA estimates that we spend on environmental protection, as a country, less than half of what we spend on clothing, one-third of what we spend on defense and medical care, one-fifth of what we spend on housing, and one-sixth of what we spend on food. Moreover, while environmental investment has been increasing over the past two decades, it is actually declining as a percentage of our total capital investment—that is, we are investing a smaller percentage of our money in the environment than we were several years ago.[63]

As a second perspective, in 1992, Congress, through the Intermodal Surface Transportation and Efficiency Act (ISTEA), agreed to federal spending of roughly $30 billion a year over five years to improve our roads, highways, and other forms of surface transport. While this spending may be justified, it is ironic that we spend fifteen times as much each year to build and restore our artificial railroads and highways as we do to protect and restore the natural riparian highways that served as our country's original, natural transportation system and that provide us with so many other economic and ecological benefits as well.

Jobs

Finally, by not investing in clean water, we are *also* losing access to potential jobs. Different sources indicate that investment of $1 billion in water and wastewater infrastructure will generate between 6,400 and 15,600 jobs directly involved in project completion. Indirectly, as many as 13,600 jobs could be created per billion dollars invested. Total effects have been estimated at 34,200 to 57,400 jobs per billion dollars invested. And the estimated $8.3 billion (1991 dollars) shortfall in funds for water and wastewater capital for the period 1993 to 2000 could represent 2,865,900 to 4,810,000 job-years of employment.[64]

According to the Maryland Institute for Ecological Economics, even more jobs can be created by investing in "natural infrastructure," through the restoration of wetlands, streambeds, fisheries habitat, and other essential components of aquatic ecosystems. These economists predicted that investments in aquatic ecosystem restoration would produce an average of thirty jobs per million dollars spent, a job creation rate 37 percent higher than generated through public investment in roads, 24 percent higher than through water and sewer systems, and 28 percent higher than through major defense contracting.[65] The National Academy of Sciences (NAS) proposes a long-range program to restore badly degraded aquatic resources.[66] Investment in such a program could generate tremendous long-term ecological as well as economic benefits.

Increased public spending to protect and restore our aquatic ecosystems, then, produces multiple public benefits. Most important, it will help to provide cleaner water and healthier habitat for fish and wildlife. This, in turn, can improve human health and restore fish and wildlife populations that are valuable for their own sake and that support important industries. Incidentally, but important, public investment in clean water also creates jobs in many sectors of the economy.

Pollution Prevention Pays

These arguments all justify increased public spending for public water quality programs—federal and state management of pollution control programs, construction of public sewage treatment plants, and restoration of aquatic ecosystems. Additional private investment will be needed as well, however, to move us further toward the zero discharge goal of the Clean Water Act and to reduce the tremendous damage caused by runoff from farms, factories, and parking lots. Private businesses that seek to profit from their activities should pay the full costs of preventing harm to public resources. Inevitably, however, arguments arise that these businesses pay a price higher than the value of the public resources they may destroy.

Given the difficulties in valuing water resources previously discussed, this debate may proceed for a long time. Fortunately, however, smart businesses are learning that the trade-off between environmental protection and profits may be a false one. Evidence is growing that, in both the industrial and agricultural sectors, sound environmental practices can help save money and make businesses more efficient and more profitable. Three studies prove the point: one involving the country's largest chemical plants, a second addressing a range of small- to medium-sized industries, and a third dealing with sustainable farms around the country.

A study released recently by INFORM, a New York–based research group, evaluated 137 source reduction activities that will reduce the use and release of toxic pollutants by twenty-one of the country's largest chemical plants. The results show that these corporate environmental improvements will have economic benefits as well. Out of seventy cases affecting product yield, sixty-eight resulted in increased production. Almost half of the actions will result in direct monetary savings, through reduced product or process costs and other savings; 15 percent of these savings will exceed $1 million per year. By contrast, only one of the activities will increase net costs. One-quarter of the actions required no capital investment, and about half cost less than $100,000, with expected paybacks (time by which resulting savings will pay back this investment) of eighteen months or less.[67]

In a similar study, Citizens for a Better Environment (CBE) found that a wide range of businesses in Silicon Valley, California, could reduce toxic chemical pollution of San Francisco Bay while saving money and jobs at the same time. Based on this study, CBE concluded that industries in the region could reduce by 75 percent to 90 percent their releases of the toxic metals nickel and copper, fingered as key toxins affecting aquatic life in the Bay, while saving money and preserving an estimated forty thousand jobs. Some firms had demonstrated this apparent miracle, and CBE concluded that others could replicate it by substituting less toxic (and less expensive) materials, changing the production process or temperature, and filtering and recycling process water to improve product purity and recover valuable metals and other materials. The capital costs needed to fund these improvements ranged from $10,000 to $1.3 million, with payback periods from three months to five years.[68]

Finally, a detailed study released in 1989 by NRC concluded that alternative agricultural practices could reduce off-farm releases of chemicals, sediments, and other sources of pollution while maintaining economically viable farms. Such savings result from reductions in expensive farm inputs (pesticides and fertilizers), increased sustained productivity through reduced erosion of valuable topsoil, and improvements in plant and animal health. Specifically, the NRC study found that "[f]armers successfully adopting [alternative, environmentally beneficial] systems generally derive significant sustained economic and environmental benefits. Wider adoption of proven alternative systems would result in even greater economic benefits to farmers and environmental gains to the nation."[69]

This is not to say that additional progress in meeting the goals of the Clean Water Act will never cost money. Some improvements will continue to require private as well as public investments that may not result in immediate economic benefits to a specific firm. Increasingly, however, smart businesses, from farms to factories, arc discovering ways to make pollution pay economic as well as environmental dividends.

Conclusion

In *Earth in the Balance,* then-Senator Al Gore quoted Oscar Wilde: "A cynic is one who knows the cost of everything and the value of nothing." It is easy to dwell on the costs of clean water, while forgetting the tremendous value of aquatic resources to our economy. We share a large stake in the protection of our rivers, lakes, and coastal waters, but we have not done an adequate job of preserving them. These failures bring tremendous economic as well as environmental harm. Whether we view our Clean Water Act record as ecologists or as economists, we must conclude that we have not succeeded in the mission we embarked on twenty years ago.

Have we not invested enough in Clean Water Act programs? Or have we simply not gotten our money's worth from the dollars we have spent? The following chapters take a critical look at each major Clean Water Act program to determine where we have gone wrong and where we have done well.

Part II
Assessing Clean Water Act Programs

In Part I, we outlined the perilous state of our aquatic resources and examined the importance of those resources to our quality of life. It is clear that something is not working —or is not working well enough. Despite the major national effort made thus far to restore and maintain our waters, many waterways remain badly polluted, and many critical habitats—habitats designated as essential to the survival of one or more species—are threatened or impaired. Human health, aquatic life, and the nation's economic health remain at serious risk.

In Part II, we explore why victory remains elusive. We evaluate the Clean Water Act programs designed by Congress to identify and resolve water quality problems and to measure our success in addressing those problems. The goal is to look beyond the Clean Water Act on paper to see how it is working in the real world.

The available data reveals that, in virtually every critical area, the national program is significantly behind where it should be. There is cause for concern regarding monitoring, establishment of meaningful measures for water quality (criteria and standards development and adoption), control of polluted runoff, permitting of dischargers, treatment of industrial wastes, elimination of uncontrolled sewage and stormwater discharges, keeping clean water clean, and enforcement.

First, the basic informational tools of the act are weak and in some cases nonexistent. These include standards against which to measure progress, monitoring to determine the sources and effects of water-body impairment, and communication of these problems to both the users and abusers of our surface waters.

Second, gaping loopholes remain in our programs to control discharges of pollutants from factories and sewage treatment plants, because we continue to "manage" or "treat" pollution at the end of the pipe rather than preventing pollution in the first place. These gaps allow the continued release of hundreds of millions of pounds of toxic and other pollutants every year.

Third, on a national basis, we have yet to implement any serious programs to attack poison runoff from farms, growing cities and suburbs, logging, mining, and other land uses—the largest remaining sources of chemical, hydrological, and, in some cases, physical water pollution. Until this immense but largely overlooked form of pollution is addressed, we have little hope of meeting the goals of the Clean Water Act.

And fourth, our narrow focus on chemical pollution has resulted in the massive ongoing degradation of physical and biological integrity of the nation's waters—rampant dredging, filling, and draining of wetlands and other waters; incessant development of floodplains and coastal habitat right up to the water's edge; and the continued alteration of natural waterways through dams, channelization, navigation projects, and similar artificial "improvements" to natural aquatic ecosystems. The law lacks comprehensive, working programs to protect and restore these aquatic ecosystems on a watershed-wide basis. We lack a working antidegradation policy to keep clean waters clean and healthy ecosystems healthy, and we lack restoration programs to heal our waters from past insults.

Chapter 4
The Need for Improved Standards, Monitoring, and Communication

Restoration and protection of aquatic resources require good information. We need sound and consistent standards by which to judge aquatic ecosystem health. We need adequate programs to measure progress and to identify the major sources of impairment. And we need consistent ways to communicate this information both to the users of aquatic resources—who have a right to know whether waters are safe for fishing, swimming, and other uses—and to those who are damaging those resources, so that problems can be fixed. Unfortunately, there are serious weaknesses in all of these areas.

Water Quality Standards

At both the federal and state levels, standards by which to identify water quality problems are weak and incomplete. Both EPA and states have issued water quality standards for only a fraction of the chemicals that pollute our waters and ignore many of the serious effects of those chemicals. Standards are absent for special types of waters, such as wetlands and lakes, and existing standards ignore impacts beyond the water column itself, such as contamination of aquatic sediments and accumulation of chemical pollutants in fish and wildlife. Perhaps most important, we are only beginning to design and implement standards to address the biological as well as the chemical integrity of our waters.

Water quality standards are used to measure success (or failure) in meeting several critical goals of the Clean Water Act—to have "fishable and swimmable" waters, "no toxics in toxic amounts," and, ultimately, "chemical, physical and biological integrity of the Nation's waters." In terms of chemicals, the ultimate goal of the law is zero discharge.

Water quality standards consist of two basic components. First, states must establish designated uses for their water bodies (contact recreation, recreational fishing, and so on) that define the uses for which water bodies must be protected. As explained in chapter 1, the act required states to adopt swimming and protection of fish and aquatic life as the minimum designated uses, and, wherever attainable, to meet those uses by 1983.[1]

Second, states are required to adopt water quality criteria that define the levels of pollutants or other conditions in the water body at which designated uses are impaired. Water quality criteria can be expressed in a number of ways. They can consist of numerical measures for pollutants ("1 milligram per liter of pollutant X") or narrative criteria ("no toxic pollutants in toxic amounts"). Criteria can also express broader measures of aquatic system health, such as overall stream integrity as measured by the abundance and diversity of aquatic species, compared to abundance and diversity in relatively undisturbed systems (sometimes called *biocriteria*). Where multiple uses are designated for a water body, the most protective criterion controls.[2] The Environmental Protection Agency provides guidance to states on levels of pollutants that pose a risk to human health and aquatic life. States may issue water quality criteria based on EPA's national guidance or on other appropriate scientific information to reflect state-specific factors.[3] The Environmental Protection Agency must adopt adequate water quality standards where a state fails to do so.[4]

Water quality standards translate into action in several ways. They are the basis for stricter controls on point sources through water-quality–based effluent limits in National Pollutant Discharge Elimination System (NPDES) permits. Under section 401 of the act, they support state water quality certifications, which determine whether other federally permitted activities, such as hydroelectric projects and forest leases, will impair water quality. Finally, they are the grounds for states to adopt and enforce antidegradation policies to ensure that clean waters remain clean, consistent with the Clean Water Act's mandate that we maintain the integrity of our waters.[5] This chapter addresses the adequacy of the water quality standards themselves; in later chapters we will evaluate the adequacy of the implementing mechanisms.

Efforts to adopt and implement effective water quality standards have been deficient in large part because of a lack of EPA leadership in establishing comprehensive water quality criteria guidance addressing the full range of impacts to health and aquatic resources, as required by the law.[6] Several aspects of this mandate warrant emphasis. First, criteria were supposed to address all identifiable effects on health and welfare. Criteria that address human health but do not address aquatic life, or cancer but not other human health effects, do not meet this mandate. Second, criteria were supposed to address all types of water bodies. Criteria that address fresh but not marine water, rivers but not lakes or wetlands, or surface water but not groundwater do not comply fully with the statute. Third, criteria were supposed to address concentration and dispersal through chemical, physical, and biological systems. Criteria that apply to the water column but fail to account for contamination of sediment or fish do not meet the statutory command. Finally, criteria were supposed to include factors necessary to restore chemical, physical, and biological integrity. Criteria that address chemical but not physical impairment, therefore, would miss the most fundamental objective of the law.

Failure to Cover Many Pollutants

A 1976 consent decree between NRDC and EPA,[7] which was incorporated into the Clean Water Act in the 1977 amendments, specified EPA's duty to promulgate numeric water quality criteria. Under this agreement, water quality criteria for 126 "priority toxics" were to be completed by December 31, 1979. The agency has issued water quality criteria in some form for 109 priority pollutants. Thus, criteria are still lacking for 17 of the priority pollutants.

More disturbing is the pace at which EPA is filling these gaps. Only twelve new toxics criteria were published between 1980 and 1986, at a rate of just over two per year. A 1991 study by the GAO found that "EPA... has been slow in developing and revising criteria documents for setting numeric limits for the 126 priority pollutants, as required by the Clean Water Act."[8] But, as GAO also noted, "EPA issued nearly all of these documents between the early and mid-1980's, and has published updates to less than one third of the human health criteria documents. According to the Chief of EPA's Criteria Branch, none of the aquatic life criteria have been revised."[9] In the past several years, although a number of new draft criteria and draft revisions to existing criteria have been published, none has been finalized.

In addition, criteria are needed for pollutants other than priority pollutants. The list of priority pollutants served an extremely useful purpose in 1976 by focusing EPA's resources on pollutants that, based on information available at that time, were most critical to protecting health and the environment. But fourteen years have brought new chemical products and new wastes, additional chemical monitoring data, and new understanding of the effects of those pollutants. As GAO noted, "[s]ome state officials believe that nonpriority toxic pollutants are causing serious water quality problems."[10]

Society uses and releases to the environment thousands of chemicals beyond those subject to water quality criteria. While pollutant-specific criteria cannot be written for all of these chemicals, the inadequacy of the existing list of 126 priority toxics is evidenced by the large number of chemicals known to be toxic but not covered by the priority pollutant list. To determine the adequacy of the *Toxics Release Inventory* (*TRI*), NRDC analyzed the range of toxic chemicals released to the environment; this analysis is equally illuminating in identifying holes in the water quality standards program. For example:[11]

- The *TRI* requires reporting on more than 300 chemicals; 230 of these are discharged to surface waters, but only 80 are subject to water quality criteria.[12]
- Under RCRA, EPA listed 368 chemicals that cause wastes to be labeled hazardous.
- The 1990 Clean Air Act named 189 hazardous air pollutants.
- For purposes of the California Safe Drinking Water and Toxic Enforcement Act of 1986 (Proposition 65), California has listed 351 chemicals known to cause cancer and 101 that cause reproductive toxicity.
- The Environmental Protection Agency's Integrated Risk Information System is a database with health risk information on more than 450 chemicals.
- All told, almost 1,100 chemicals are listed as toxic or hazardous on fourteen separate federal and state chemical lists (deleting all duplications).

But the absence of criteria for industrial toxics is not the only major gap. Different types of criteria are needed to measure the impact of diffuse runoff sources, such as eutrophication from excessive nutrients. The GAO found that EPA had not developed criteria documents or comparable technical information for states to use to develop water quality criteria for controlling polluted runoff.[13] A related example is the lack of water quality criteria for a wide range of toxic pesticides currently in common use. Pesticides on the priority pollutant list are those that were used widely in or before the 1970s; some of them are no longer in use at all.

Failure to Address the Full Range of Human Health and Environmental Effects

Even EPA's existing water quality criteria do not cover the full range of human health and environmental effects of many pollutants. Some address human health but not aquatic toxicity; freshwater but not marine toxicity, or vice versa; or acute but not chronic toxicity, or vice versa. As of July 1991, EPA had issued 108 human health and 22 aquatic life criteria documents for priority pollutants, and 9 human health and 10 aquatic life criteria documents for nonpriority pollutants.

Typically, EPA establishes criteria based on the most sensitive human health or environmental risk. This approach would be acceptable under three conditions: (1) if the health or environmental effect that forms the basis of a criterion clearly represented the most sensitive end point; (2) if these criteria represented mandatory minima, that is, if states could only promulgate criteria at least as strict as the most sensitive EPA criteria; and (3) if these criteria were always applied conservatively.

These conditions are not always met, however, as indicated by the recent controversy over dioxin (2,3,7,8-TCDD). The Environmental Protection Agency recommends a criterion of zero to achieve complete protection, based on the assumption that dioxin at any level may cause cancer. But neither EPA nor the states take this recommendation seriously. Instead, EPA presents potential criteria to address lifetime cancer health risks of one in one hundred thousand to one in 1 million. (About 20 to 25 percent of the U.S. population will die of cancer. A one in 1 million incremental cancer risk means that out of every million individuals exposed to the assumed levels of the pollutant for seventy years, one additional person is estimated to get cancer.)

While the criteria document and other EPA documents indicate that other human health effects may occur at slightly higher levels, no actual numeric criteria have yet been developed for human health effects such as reproductive toxicity and liver damage. Thus, when some states elected to issue dioxin criteria a hundred times weaker than EPA's criterion, they did not consider whether they jumped over levels at which other serious health effects occur. So even if the weaker standards adequately address cancer risk (a highly questionable assumption), we may be putting our unborn children at risk.

A similar situation exists with respect to aquatic toxicity. Chronic effects to fish and other aquatic life may occur at dioxin levels only slightly higher than EPA's recommended criterion to protect people against a one in 1 million cancer risk, EPA reported recently,[14] and at levels considerably lower than some state dioxin standards based only on cancer risk, with no consideration of aquatic toxicity.[15]

Faced with massive pressure on the issue, EPA is in the process of a comprehensive reassessment of dioxin. But no similar efforts are under way for most other pollutants. The bottom line is that EPA is legally obligated to consider all identifiable human health and environmental effects but has not done so for many pollutants.

Failure to Reflect the Latest Scientific Information

Most of EPA's water quality criteria are a decade or more old. With respect to many criteria, information on health and environmental effects may not have changed significantly. This is not the case, however, with respect to some pollutants. Two examples, one specific and one generic, demonstrate this point.

Returning to dioxin, EPA's cancer risk analysis is based, in part, on an assumed bioconcentration factor (BCF) of 5,000, meaning that dioxin reaches concentrations in fish 5,000 times higher than in the water. Recent evidence shows BCF levels for dioxin more than thirty times higher.[16] And as discussed in chapter 2, bioconcentration alone does not account for food chain uptake. Use of BCF alone can underestimate actual exposure to toxic chemicals by a factor of 100.[17]

A more far-reaching example is EPA's use of an assumed average human fish consumption rate of 6.5 grams per day for its risk assessments for all non-threshold carcinogens. As a preliminary matter, EPA is legally obligated to protect subpopulations that consume higher than average amounts of fish, such as recreational and subsistence fishers. Equally important, EPA's assumption is based on survey data that are more than a decade old.[18] More recent data indicate significantly higher consumption rates, particularly by certain subpopulations, such as recreational and subsistence fishers, and certain racial and ethnic groups.[19]

To compound these problems, EPA's existing criteria assume that people, fish, and wildlife are exposed only to individual pollutants, from water sources alone. The criteria ignore simultaneous, perhaps additive or synergistic effects from multiple pollutants and multiple sources (air, water, food, and so on).[20]

Failure to Address a Range of Water Bodies

EPA has a long way to go in issuing water quality criteria for freshwater and marine systems. But inland rivers and the open ocean do not cover the range of aquatic ecosystems. Systems such as wetlands, estuaries, and lakes need special consideration. Lakes and wetlands, for example, typically exhibit far longer retention times than flowing rivers and may demand stricter criteria on persistent toxics—in many cases, zero.[21] The high productivity and different temperature and salinity conditions in estuaries similarly require special consideration in the issuance of water quality criteria. In most respects, EPA has not issued special water quality criteria for these systems.

Failure to Address Contamination of Sediment, Fish, and Aquatic Life

A glaring problem with EPA's water quality standards is their focus primarily on the concentration of pollutants in the water column. This approach only partially takes into account the statutory command that EPA consider the concentration and dispersal of pollutants, or their by-products, through biological, physical, and chemical processes. By ignoring or partially ignoring contamination of sediment and fish and shellfish, EPA fails to protect against the full range of health and environmental impacts of toxic pollutants. Instead, EPA improperly relies on the water itself to dilute dangerous pollutants.

To begin, an exclusive focus on water column concentration assumes, at least for the most part, that toxic pollutants remain dissolved or suspended in the water. This allows the discharge of an extremely large mass of toxic pollutants so long as the concentration of the effluent is sufficiently low. This is problematic particularly for discharges of runoff, which occur during high-flow (and therefore high-dilution) conditions, and for large-volume discharges. Toxic pollutants do not all remain in the water column; many are sediment-bound rather than soluble and, as shown in chapter 2, can accumulate in the sediment in dangerous amounts. Enforceable sediment quality criteria are needed to address this problem. While EPA is moving slowly to issue sediment criteria, no final action is expected in the near future.

Pollutants also accumulate in fish and other wildlife. In theory, the development of ambient water column criteria takes this factor into account. But as discussed in the context of the BCF for dioxin, our understanding of bioaccumulation (bioconcentration and biomagnification) is incomplete at best. Moreover, bioaccumulation is considered largely to address human health effects from consuming contaminated fish and shellfish. Omitted from the analysis are acute and chronic effects on wildlife, including not only fish and aquatic life but birds, mammals, and other species that consume contaminated aquatic life or otherwise are exposed to toxics in the aquatic environment.[22] Criteria governing the presence of toxics in aquatic organisms would provide a second line of defense, but EPA has issued no criteria to address these impacts.

Just Beginning to Address Biological Impairment

More than ten years ago, FWS and EPA realized that standard water quality monitoring efforts did not provide complete, meaningful information about the biological health of surface waters:

> Until recently, attempts to monitor the condition of the Nation's waters have focused only on the physical and chemical characteristics of the water, while the components of the biological communities were largely ignored. Additionally, these physical and chemical data were not collected in the context of a statistically designed evaluation. A link between the physical and chemical characteristics and the associated health of the biological communities was clearly needed.[23]

Only in the past several years, however, has EPA moved seriously to address these problems. Now EPA and many states are beginning to use biological water quality criteria, which measure the overall health of a water body by comparing the abundance and diversity of a representative range of species there to abundance and diversity in control waters that remain in a relatively natural state (discussed in more detail in chapter 2). According to EPA, "[t]his indicator is the most direct measure possible of support of a Clean Water Act (CWA) goal, because maintaining biological integrity is one of the legislative mandates."[24]

Biological criteria supplement rather than supplant other water quality criteria. In 1990, EPA issued national guidance indicating that use impairment should be based on any one of three types of data—chemical, biological, or toxicological impairment (regardless of findings of other data).[25] The agency believes that "[r]ecent studies indicate that integrated assessments, including biological, chemical and physical data, convey a much more accurate and complete picture of water resource status than an approach based solely on chemical criteria."[26]

While some states use biological monitoring in at least a few water bodies, however, and some have incorporated these procedures into regular monitoring networks, most use this process for only a relatively small percentage of waters.[27] As of April 1990, when EPA's national biocriteria guidance was issued, only twenty states were using some form of ambient biological assessments, although the level of effort and sophistication varied considerably. Only fifteen states were using this information to develop actual biological water quality criteria. Because EPA has not required states to adopt biocriteria as a part of their water quality standards program, it is likely to be several years before biological criteria become a full partner in EPA and state water quality programs.

State Water Quality Standards Even Further Behind

EPA's preparation of water quality criteria documents (which are not legally binding) is just the beginning of the process; states must adopt these (or other scientifically sound) criteria for themselves in order to keep the monitoring and permitting programs on track. Slow state action in adopting numerical criteria for toxic pollutants means slow state action in translating these criteria into meaningful permit limits or water quality certifications.

Concerned that states had not moved rapidly enough in the area of toxic pollutants, Congress directed in 1987 amendments to the Clean Water Act that states adopt within three years (by 1990) numerical criteria for all priority toxic pollutants "the discharge or presence of which" could be expected to cause water quality problems.[28] Many States were slow to respond.[29] As EPA itself acknowledged in 1991, "[a]vailable standards against which POTW discharges are judged (and limited) are developed inconsistently across States.... Significant nationwide reductions in toxic discharges can be expected once appropriate standards for receiving media are developed and implemented consistently among States."[30]

In April 1990, EPA identified twenty-two states and territories that had failed to comply with the 1987 requirement that they adopt water quality standards for toxic pollutants.[31] Despite this notice, a number of states continued to delay. In November 1991, EPA formally announced its intent to issue criteria in the twenty-two states (and territories) that had not yet acted.[32] It took EPA until December 1992 (and a lawsuit by NRDC),[33] however, to issue a final regulation setting water quality standards for the errant states.[34] And even this action had a bitter twist. Rather than issuing consistent criteria for all states, EPA chose different cancer risk levels for different states, based on EPA's speculation about what risk level the state would have chosen if it had complied with the law. The result: Citizens in some states will face cancer risks ten times higher than the public in other areas.

This inconsistency is just one example of sometimes wide disparities in the stringency of water quality standards. Because EPA has given states wide latitude to diverge from its water quality criteria guidelines, state criteria often differ greatly.[35] While differences that reflect regional ecological conditions may be justified, varying levels of public health protection are not.

Failure to Ensure Protection of Endangered Species

Requirements of other federal laws, such as the National Environmental Policy Act, supplement Clean Water Act provisions designed to protect biological integrity. While the relationship between these laws and the Clean Water Act is far beyond the scope of this book, one statutory mandate is particularly important: section 7 of the Endangered Species Act (ESA).

The ESA is the principal federal statute designed to protect endangered and threatened species and the ecosystems on which they rely for survival. To accomplish this goal, section 7 of the ESA casts a broad net. It requires all federal agencies to consult with the secretary of the interior (implemented in practice through the director of FWS) to "insure that any action authorized, funded or carried out by such agency... is not likely to jeopardize the continued existence of any endangered species or threatened species or result in the destruction or adverse modification of [critical] habitat of such species."[36]

Until recently, EPA has taken no major steps to honor this consultation requirement in its implementation of the Clean Water Act. Given the large number of threatened, endangered, or other imperiled species of fish and other aquatic life discussed in chapter 2, EPA compliance with the ESA appears long overdue.

In a recent lawsuit that highlighted EPA's noncompliance with the ESA, an Alabama citizen challenged the agency's failure to comply with section 7 of the ESA in approving a number of aspects of Alabama's water quality program, including the review and issuance of water quality standards, state grant funding under section 106 of the CWA, and Alabama's issuance of NPDES permits (or EPA's review of such permits). An FWS official in Alabama first flagged the issue with a letter to the Alabama Department of Environmental Management stating that proposed changes to the state's water quality standards may affect several species listed or proposed to be listed by FWS as threatened or endangered.[37]

The Alabama lawsuit has not yet been resolved. In part in response to the claims raised in that case, however, and in part based on negotiations between EPA and the Interior Department dating to 1990, in July 1992, EPA, the Department of the Interior, and NMFS signed a memorandum of understanding governing EPA compliance with section 7 in the water quality standards program. Under this memorandum, EPA will

● consult with FWS and NMFS in the review of existing EPA water quality criteria and in the issuance of new water quality criteria regarding the impact of aquatic life criteria on threatened and endangered species;

● conduct a broad, national assessment of the adequacy of existing water quality standards to protect threatened and endangered species;

● engage in informal consultation with FWS and NMFS to determine which species may be affected by aquatic life water quality criteria and to prepare a biological assessment of those effects; and

● initiate formal consultation with FWS and NMFS for species that are likely to be adversely affected by existing criteria and, presumably, revise the criteria to address those effects.[38]

The Memorandum of Understanding is a major step forward in the effort to ensure that water quality standards protect threatened and endangered species and EPA's first major assumption of its duty to comply with the ESA in its implementation of the Clean Water Act. Serious questions remain, however, about its adequacy. By consulting with FWS and NMFS only when EPA's water quality criteria for aquatic life are issued under section 304(a), EPA ignores the critical step of assuring that state water quality standards and the programs designed to implement them—including, for example, NPDES permits and antidegradation programs—translate EPA criteria into actual protection of endangered species and their habitat.

Inadequate Monitoring and Public Information

Inadequate and inconsistent monitoring at all levels hamper federal and state water quality efforts. Only a small fraction of waters are tested routinely for toxic and other pollutants, and the data that do exist are often inconsistent. Few states monitor their waters for broader biological impacts, such as loss of habitat and biological diversity due to physical changes. Even compliance monitoring, which accounts for the bulk of existing efforts, provides incomplete and inconsistent data about program effectiveness, or to target future monitoring, enforcement, or other actions. Most alarming is the absence of comprehensive, consistent programs to detect and warn the public about serious health threats from contaminated swimming and fishing waters.

Monitoring is essential to effective implementation of the Clean Water Act. Without adequate monitoring, it is impossible to assess trends or evaluate whether prevention or control programs have worked. A recurrent theme in chapter 2 was the inadequacy of current efforts to gather and assimilate data on water quality and aquatic resource trends. The history of the national water quality monitoring program is one of gaps, overlaps, and inadequacies, although coherence and logic are not being brought into the effort.

Ambient Water Quality and Compliance Monitoring

Twenty years after the Clean Water Act mandated that the states monitor their waters and report their findings to EPA and Congress, only a fraction of state waters have been assessed regularly and meaningfully.

The percentage of waters assessed varies wildly among states. According to EPA's 1988 *National Water Quality Inventory*, four of the forty-eight states reporting assessed the quality of 10 percent or less of their rivers and streams, whereas only ten states assessed most of these waters.[39] And rarely is actual sampling done. A recent GAO study of state monitoring programs reported that, "[a]ccording to the 1988 Water Quality Inventory, only forty percent of assessed rivers and streams were actually sampled to determine water quality; the remaining sixty percent of assessed waters were evaluated by using descriptive data."[40] Most EPA and state officials agreed that ambient monitoring for toxics is essential to determine the extent and sources of toxic pollution. Yet three out of the four states visited by GAO as part of its survey did no routine ambient monitoring for toxics.[41] While the last several 305(b) reports show that monitoring efforts are improving in general,[42] we are still a long way from having a complete monitoring program (see Table 4.1).

The lack of systematic state monitoring is due in part to the absence of enforceable EPA requirements that specify monitoring methods or frequency

TABLE 4.1

Percentage of Waters Monitored by State

	STATES REPORTING	% ASSESSED	% MONITORED
Rivers			
1990	51	63	37
1988	48	45	40
1986	42	21	—
Lakes			
1990	46	69	26
1988	40	73	—
1986	37	32	—
Estuaries			
1990	22	66	34
1988	23	76	—
1986	20	55	—

Sources: EPA, National Water Quality Inventories (1986, 1988, 1990).

to identify toxic and other pollutants, and other impacts from point and nonpoint sources. The quality of monitoring also varies from state to state, which contributes to additional uncertainty.[43] The Environmental Protection Agency does have regulations requiring states to have water quality monitoring programs, but these requirements are vague and cursory, leaving the design of monitoring programs primarily to individual states.[44] For example, EPA's rules do not specify how many waters must be monitored, for what pollutants, and at what frequency.

Monitoring is expensive and must be well coordinated among widely dispersed locations to be effective. One constant, however, is the problem of resource constraints that hamper state monitoring; of the four states GAO visited for its survey, three (Georgia, Michigan, and Massachusetts) reported that they were planning to or already had rolled back their limited monitoring programs due to budget cuts.[45]

Another reason for the small number of stream miles assessed is that the vast majority of the samples taken from our waters are for purposes of monitoring compliance by individual pollution sources rather than for ambient water quality assessment.[46] In other words, permittees or government officials test water quality downstream of discharges to see if discharge limits are being met. A 1990 NRC report found that the U.S. spends more than $130 million annually on coastal environmental monitoring but that most of these resources have been devoted to compliance monitoring, which by nature focuses on small areas over a short time.[47]

Even in the area of point source compliance monitoring, however, efforts are weak and inconsistent. In 1990, EPA admitted that the availability and quantity of point source data is limited and that most of the data that are collected are for flow and conventional pollutants, with relatively little information on toxic metals and organics.[48] This phenomenon may result from the overriding reliance on self-reporting, the system of discharge monitoring reports required to be submitted by dischargers under section 308 of the Clean Water Act. This information is extremely useful to support government and citizen enforcement efforts, but few of the data are analyzed; most simply are filed away in government records and databases.

One example is the lack of comprehensive information on the effectiveness of the National Pretreatment Program, which is intended to prevent toxic pollutants from being discharged to sewage treatment plants in amounts that will pass through the plants and contaminate U.S. waters. Despite millions of dollars spent on controlling toxics through pretreatment, we still cannot fully answer the question: Is the program working? A recent EPA report on the pretreatment program concluded:

> [F]ew data currently exist on which to base a true evaluation of environmental impacts due to discharges of toxic pollutants by POTWs. Existing national data bases are limited in scope and completeness, and often are not quality controlled sufficiently for rigorous analysis. Other studies and data collection efforts are too fragmented to contribute to a consistent and reasoned view of impacts caused by toxic discharges from POTWs.[49]

These problems are consistent with regional analysis. A review by the Chesapeake Bay Foundation (CBF) of 160 sewage treatment plants in Maryland, Pennsylvania, and Virginia found that only one plant in Maryland

and thirteen in Pennsylvania reported regularly on releases of toxic metals. More plants, but still less than half, had requirements to measure whole efflu- ent toxicity.[50]

As discussed previously in this chapter and in chapter 2, EPA has made signif- icant efforts in recent years to strengthen federal and state monitoring pro- grams and to make them more consistent. The agency has engaged in a com- prehensive review of monitoring efforts[51] and is taking steps to implement needed improvements, most recently in cooperation with other federal or state agencies.[52] But we still have a long way to go to improve monitoring to identify and assess progress in correcting remaining sources of water pollution.

A related problem is the lack of easy public access to compliance informa- tion. Section 308 of the Clean Water Act provides citizens with an absolute right to effluent data collected from any source, including discharger self-monitoring data. Since section 101(e) gives the public the right to participate in all permit- ting and other proceedings under the act, most basic program documents are legally available to the public. Indeed, this information is essential for citizens to exercise their right to bring citizen suits against polluters or government offi- cials under section 505 of the act.

When it comes to access to information, however, there is a big difference between rights and reality. Public access to reports and records is often limited to remote state offices, and information is often difficult to locate. Reports sometimes are filed months after they are submitted, and citizens may have to wait weeks for an appointment to review the files. Nor is there any easy, public way for citizens to know who is discharging pollutants to which waters and where they can get copies of the permits that allow these releases. While much of this information is computerized on EPA's Permit Compliance System data- base, this database is not currently available to the public in a user-friendly form.

Monitoring and Reporting to Protect Public Health

It is bad enough that our overall monitoring efforts are inadequate for pur- poses of detecting water quality trends and weighing the effectiveness of clean water programs. The public should at least be able to expect, however, ade- quate monitoring of swimming and fishing waters to detect severe public health threats from biological and chemical contamination, and appropriate procedures to warn the public about these threats.

In a 1991 study of monitoring and closure practices at U.S. coastal beaches, NRDC found that many states do not monitor coastal beach water quality regu- larly. To the extent that they did test for disease-bearing organisms, many states did not follow EPA's recommended guidelines for indicator organisms or use the recommended testing protocols—making the results less accurate or meaningful in determining human health risks. The report found that ten states monitor their ocean beach waters infrequently if ever; eight others monitor portions of their coasts; and only four monitor their whole coasts.[53] (The report's findings regarding the numbers of closed or polluted beaches were presented in chapter 2.)

Similar problems exist in the way information about polluted beaches is used even when problems are identified.[54] Some beaches where high (and potentially risky) levels of disease-bearing organisms were found still were not closed, nor were warnings posted for bathers. Thus, along many U.S. beaches today, people are bathing at their own risk. Only Connecticut, New Jersey, and Delaware consistently close beaches every time water quality standards for bacteria are violated. Some states issue prompt warnings, while others wait weeks or even months to publicize unsafe swimming conditions.

Similarly, there are serious weaknesses and inconsistencies in state monitoring and reporting of fish contamination. A 1989 EPA survey of state fish consumption advisory practices[55] showed extreme variations among states in the methods used to evaluate human health risks and to report that information to the public. For example:

- Only twenty-nine states had routine monitoring programs to identify fish contamination, while eleven monitored in response to emergencies and thirty-seven relied on special studies of particular waters.
- States varied in what portions of the fish they measured for toxics—whole fish (sixteen states), fillet with skin (twenty-three), skinned fillet (twenty-three). (Note: Some states use multiple methods.)
- States measured different pollutants and used different testing procedures.
- Thirty-four states based bans and advisories on FDA action levels; others used EPA risk assessment procedures; and others used a combination of the two.
- States made different assumptions to calculate how much health risk results from a given level of contamination. Varying assumptions included consumption rates, acceptable risk levels, average body weight, and meal size.
- States differed in whether they issued consumption advisories indefinitely or for fixed time periods and in the length of these periods.

As discussed in chapter 2, these procedures likely underestimate human health risks. In establishing procedures, states generally rely on outdated FDA action levels for a handful of pollutants, fail to address the cumulative effects of multiple pollutants, and use consumption levels appropriate only to average fish consumers, leaving subsistence and recreational fishers at higher risk. Equally disturbing is the variation in the amount of effort by different states to identify seafood contamination problems. Even in the best states, only a tiny fraction of waters are monitored routinely for toxic pollutants in fish (see Figure 4.1).

Similar problems and inconsistencies exist with respect to reporting of fish contamination. There are no national posting or reporting requirements for recreational fishing areas. EPA's survey found that states varied widely in the methods used to announce fishing bans or advisories, with options including signs posted at water bodies, information distributed in fishing licenses, announcements on television or radio, and notices in newspapers or in public

FIGURE 4.1

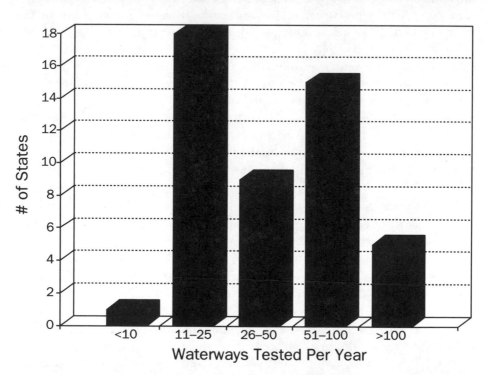

Source: Patricia A. Cunningham, Ph.D., J. Michael McCarthy, and Devorah Zeitlin, *Results of the 1989 Census of State Fish/Shellfish Consumption Advisory Programs,* prepared for EPA Assessment and Watershed Protection Division, Office of Water Regulations and Standards (Research Triangle Institute, 1990).

buildings. (The complex mix of systems to address testing and regulation of commercial fisheries is beyond the scope of this analysis.)

A 1989 NOAA survey of state-issued fish consumption advisories reached similar conclusions:

> The number of fish advisories issued by each of the coastal and Great Lakes states... show[s] great variability. These differences do not necessarily reflect the degree of pollution in fish tissue but more likely reflect the effectiveness of state systems in identifying potential health risks and issuing advisories. Budgetary constraints often limit the state capacity for sampling and monitoring toxic substances in fish populations.[56]

NOAA agreed that sampling programs are limited in scope; monitoring and reporting techniques vary among, and even within, states; and assumptions about exposure and consumption and risk assessment models all vary widely.[57]

The inequities that may result from these differences in state procedures are best understood by way of example. The same concentrations of PCBs from nine hypothetical trout taken from the Great Lakes, according to Dr. Jeffrey Foran of George Washington University, would result in vastly different health advisories in different areas. Consumers would be advised not to eat any of the fish taken from Lake Michigan, Lake Huron, or parts of Lake Superior. Most would be advised to eat the fish from Lake Ontario no more than once a month, while pregnant women, women of child-bearing age, and children would be advised not to eat Lake Ontario fish at all. But consumers would be told that the fish from Lake Erie were safe in any amount and for any person.[58]

To make matters worse, the absence of a clear, consistent national policy to determine whether swimming beaches and fishing waters are safe is coupled with the absence of a clear link between these health advisory systems and the Clean Water Act regulatory process. The public expects not only fair warning of unsafe beaches and seafood, but action to reduce and ultimately eliminate these risks. Depending on the location, however, a given level of contamination could result in no warnings, warnings but no regulatory action to correct the problem, or warnings plus additional measures to correct the problem.

Clearly, with respect to both swimming and fishing, the United States lacks a coherent nationwide system for informing the public when the waters they use are safe. In general, warnings are not posted for bathers and anglers at water bodies that violate water quality standards or that have tainted fish, and there are serious differences in whether or how states warn the public about these risks.

Chapter 5
The Elusive Zero Discharge Goal

When most people think of water pollution, they picture noxious chemicals spewing from a factory pipe into a nearby river or lake. Many people even realize that the wastes from their sinks and toilets may go to a public sewage treatment plant (publicly owned treatment work, or POTW), whose wastewaters also are released into a local waterway. The federal and state governments have largely agreed with this stereotype and have focused water pollution controls over the past twenty years on these so-called point sources.[1]

In conspicuous defiance of the Clean Water Act's 1985 zero discharge goal, however, a massive number of point sources continue to discharge wastes into the nation's waters. Together, as of November 1991 EPA and the states had issued about 65,000 permits to release pollutants into surface waters in the United States. These include about 21,000 to POTWs and 43,400 to industrial or commercial facilities.[2] In addition, an estimated hundreds of thousands of municipal and industrial stormwater pipes release polluted runoff into the nation's waters; most of these are just beginning to receive permits. This chapter evaluates the adequacy of our efforts to control pollution from these point sources.

Regulatory Structure for Industry and Sewage Treatment Plants

As discussed briefly in chapter 1, in the years preceding the 1972 Clean Water Act, water pollution was regulated through a scheme whereby each state decided upon the uses of its waters, established water quality criteria that supported those uses, and then regulated, on an individual basis, pollution sources that caused water quality to exceed those criteria. This approach resulted in little water pollution control.

In 1972, Congress rejected this singular focus on the quality of receiving waters and added a nationwide technological approach to water pollution control. This shift represented an important change in outlook. The "use of any river, lake, stream or ocean as a waste treatment system" was deemed to be unacceptable.[3] Hence, the zero discharge goal—that "the discharge of pollutants into the navigable waters be eliminated by 1985."[4]

Under the technology-based approach, EPA would establish national effluent guidelines for industrial discharges, which would become increasingly stringent as pollution prevention and treatment technology advanced. Discharge elimination guidelines were to be adopted by EPA wherever feasible, particularly for new sources where the cost of innovation was expected to be lower.[5]

For POTWs, a national standard of *secondary treatment* was established.[6] Industrial facilities discharging wastes to POTWs would be required to comply with pretreatment standards to prevent pollutants from passing through the POTW into receiving waters or interfering with POTW operation.[7]

Effluent guidelines were to be implemented through a permitting program, administered by EPA or by individual states approved by EPA to assume delegated responsibility for the program.[8] Without such a permit, dubbed a National Pollutant Discharge Elimination System (NPDES) permit, the discharge of any pollutant would be illegal.[9]

Industries discharging wastes to POTWs, so-called indirect dischargers, were required to comply with pretreatment standards directly and do not need NPDES permits. The POTW itself, as the direct discharger, was required to have an NPDES permit and to adopt mechanisms to enforce pretreatment standards against industrial dischargers. Where national pretreatment standards were not yet available, the POTW was to develop local limits to control industrial discharges to the plant.[10]

Water quality standards remained as a supplemental control strategy, but only as a rest stop on the road to zero discharge. All discharges must meet national effluent guidelines regardless of the quality of the receiving water. If compliance with effluent guidelines is not adequate to prevent violations of water quality standards, however, permit limits must be made more stringent.[11]

For this point source control system to work, a number of elements (in addition to the water quality standards discussed previously) need to be operating properly: EPA must develop industrial effluent guidelines and pretreatment standards and must move us relentlessly toward the goal of zero discharge; permits must be written in accordance with the strictures of the law; and enforcement against violators must be swift and sure. Unfortunately, there are problems with all three of these components of our point source controls.

National Treatment Standards Limited and Outdated

At the center of the Clean Water Act's program to reduce or eliminate industrial discharges are national treatment standards developed by EPA for each category of industry. But EPA has lagged seriously behind in the issuance of these standards for a wide range of industries. Many industries are governed by no national regulations at all, leaving them subject only to weak, inconsistent, site-by-site permits. Other regulations are a decade or more old and do not reflect recent improvements in pollution prevention methods.

Standards for Direct Discharges to Surface Waters

For direct dischargers, NPDES permits are supposed to be based upon effluent guidelines that define the level of control that can be achieved by the Best Available Technology (BAT) for toxic pollutants and the Best Conventional Technology (BCT) for conventional pollutants, such as suspended solids or oil and grease.[12] The BAT and BCT standards are to be reviewed periodically by EPA and strengthened (made more stringent) as technology evolves. Wherever possible— that is, where technology exists that is economically achievable—the BAT standards must require the elimination of discharges altogether.[13]

Congress intended BAT standards to be the engines that drive the technology-forcing advances of the Clean Water Act, becoming ever more advanced until pollutant elimination is achieved. Congress gave EPA (or delegated states) the authority to step in and write a technology-based permit to fill the gap until these standards could be written and implemented.[14] Where EPA has not yet issued a national effluent guideline, the EPA or state permit-writer must develop a site-specific BAT standard based upon what is called Best Professional Judgment, or BPJ.

Quite simply, BAT-based permits are easier and cheaper to write than those based on BPJ (or, for that matter, than a permit based on water quality standards). In addition, a BAT-based permit tends to be more stringent than a permit written based on BPJ. This makes sense, since the BAT-based permit relies upon a major investigation by EPA engineers and economists into the performance of all wastewater prevention and treatment technologies available to an industry. Industry dischargers are far more reluctant to challenge a permit-writer who is backed up by the strength of a BAT standard than one who is relying strictly on his or her own judgment.[15]

Full compliance with BAT-based permits in the categories for industries in which such standards have been written, EPA estimates, will result in removal of about 97 percent of the priority pollutants present in raw (untreated) wastestreams.[16] But a great deal remains to be done, despite estimates in chapter 2 that current controls have resulted in the removal of about 1 billion pounds per year of toxics that otherwise would enter surface waters. Rough estimates taken from the *TRI* show that, in 1990, industrial facilities discharged

197 million pounds of toxic and hazardous pollutants directly to the nation's waterways. It is interesting that a handful of industries accounted for the vast majority of these reported releases (see Figure 5.1).

Furthermore, these figures may represent only the tip of the iceberg, as actual toxic releases far outstrip the amounts reported in the TRI. A recent report by the National Environmental Law Center, *Toxic Truth and Consequences: The Magnitude of and the Problems Resulting From America's Use of Toxic Chemicals,* estimates that at least 350 billion pounds of toxic chemicals were produced or used in the United States in 1988—more than fifty-six times the amounts reported released or transferred as wastes in the *TRI.*[18] The Office of Technology Assessment estimated that the *TRI* database reflects less than 1

FIGURE 5.1

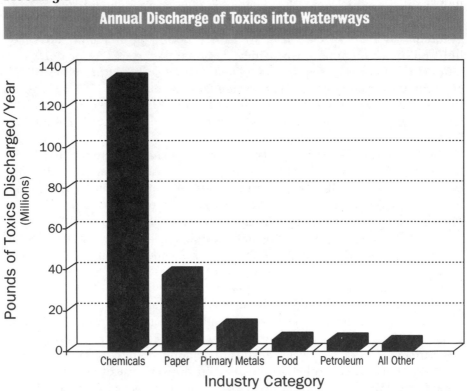

Annual Discharge of Toxics into Waterways

Source: EPA, 1990 Toxics Release Inventory, 58.[17]

percent of the actual quantities of toxic pollutants released into the environment; OTA estimated the total at 400 billion pounds.[19]

For example, because *TRI* covers only manufacturing industries, it does not cover the massive releases of toxic pollutants from oil and gas wells. Offshore drilling operations generate huge quantities of waste materials, most notably *muds* (which lubricate the drill bit) and *cuttings* (pieces of rock ground up by the drill bit that are coated with mud). Each well drilled offshore generates an average of 1,500 to 2,000 tons of muds and cuttings, which are usually discharged into surrounding waters on the outer continental shelf (OCS).[20] The Department of Interior estimated that six thousand wells will be drilled on the OCS as a result of the current five-year program.[21] Each well generates more than 7,000 pounds of priority toxics and organics, according to EPA estimates,[22] which means that more than 42 million pounds of priority pollutants will be discharged into the oceans as a result of offshore drilling activities.

The EPA effluent guidelines program has not kept pace with the scope of industries needing standards or with changes in technology since many of the existing guidelines first were written. Today, twenty years after the Clean Water Act was passed, at most only about one in three (and possibly only one in five) industrial dischargers is subject to a national BAT standard.[23] For all other dischargers that have permits, the permits are based solely on BPJ, water quality standards, or both. Many of the industries for which EPA has failed to write BAT standards are major sources of toxic discharges to our nation's waters. For example, the following industries are subject to no categorical standards:

● Hazardous waste treatment
● Machine manufacturing and rebuilding
● Transportation equipment cleaning
● Industrial laundries
● Commercial solvent recyclers
● Drum reconditioners
● Hospitals
● Used oil reclaimers and rerefiners

Other industries are subject to standards that are more than a decade old, have never been reviewed and revised, and therefore have not reflected improvements in pollution prevention or control. The most problematic industries among these each are allowed by existing BAT rules to discharge at least 1 million pounds of toxic pollutants per year (See Table 5.1).

The pace of EPA activity in writing BAT standards has slowed markedly. Although BAT standards have been written for about fifty-one industry categories,[25] EPA has promulgated only two major new effluent guidelines since 1986 (for the Organic Chemicals, Plastics, and Synthetic Fibers category in

TABLE 5.1

Industries Discharging at Least 1 Million Pounds of Toxic Pollutants per Year	
INDUSTRY	**YEAR GUIDELINE ISSUED**
Coal mining	1982
Electric power plants	1982
Iron and steel	1982
Leather tanning	1982
Pesticide manufacturers/packagers	1976
Pulp-and-paper manufacturers	1982
Petroleum refiners	1982
Pharmaceutical manufacturers	1983
Textile mills	1982
Timber products	1981[24]

Source: EPA, "Summary of Effluent Characteristics and Guidelines for Selected Industrial Point Source Categories: Industry Status Sheets," (1986).

November 1987 and for the Offshore Oil and Gas drilling category in January 1993).[26]

Congress has been well aware of the slowdown in the BAT program and of the consequences for water pollution prevention and control. In 1987, as part of a comprehensive program to improve the control of toxic pollutants,[27] it directed EPA to develop a plan to review and revise outdated guidelines and to issue guidelines for industries not covered at all.[28] The 1987 amendments gave EPA until February 1988 to publish a final plan for promulgation of these new and revised standards.

The first plan issued by EPA—and published nearly two years after the February 1988 deadline[29]—called for little more than business as usual. The

Natural Resources Defense Council challenged EPA's plan in court, resulting in a consent decree in which EPA committed to propose and promulgate more than twenty new and revised effluent guidelines over the next decade.[30]

A more fundamental problem with EPA's effluent guidelines program, however, is its general failure to adopt the zero discharge approach mandated by the Clean Water Act. While terms such as *pollution prevention* and *source reduction* are in vogue in environmental circles in the 1990s, in fact these concepts were embedded in the effluent guidelines provisions of the 1972 Clean Water Act. Best Available Technology standards were to "result in reasonable further progress toward the national goal of eliminating the discharge of all pollutants...[and] require the elimination of discharges if...such elimination is technologically and economically feasible."[31]

The Environmental Protection Agency's effluent guidelines program reflects an end-of-pipe focus, that is, an exaltation of treatment systems that reduce pollutants after they are generated over methods to eliminate the creation of pollutants in the first place. A major thrust of criticism is that end-of-pipe treatment often simply shifts pollutants around from one place to another. In the case of water pollution control, removing volatile organic pollutants from effluent often means "stripping" (evaporating) them into the air, where they jeopardize the health of plant workers and the public; and metals and other organics are shifted into industrial sludges, which must be "treated" again through incineration or landfilling, with yet more environmental risks.

The original Clean Water Act embodied the philosophy of pollution prevention, requiring EPA to consider basic process changes as well as end-of-pipe treatment methods and to evaluate the effects of various control options on "non-water quality environmental impact[s]."[32] Progression from BPT to BAT and the requirement that EPA continuously review and evaluate guidelines to reflect technological progress in pollution prevention and control reflected an intent that we move toward the national zero discharge goal.

In this respect, EPA's record is abysmal. After two decades, the agency has prescribed zero discharge requirements for only a small handful of industrial categories.[33] Without a careful review of each guideline, it is impossible to indicate each zero discharge option that has been missed. Some analysts, however, such as Professor Oliver Houck of Tulane Law School, have commented on the backward trend of weaker rather than stricter effluent guidelines in recent years.[34]

At least one example shows in two respects how EPA intentionally failed to take advantage of pollution prevention opportunities in the effluent guidelines for one of the largest sources of toxic pollution—the organic chemicals, plastics, and synthetic fibers (OCPSF) industry. First, EPA recognized that it could guarantee the simultaneous prevention of air and water pollution by requiring its OCPSF effluent guidelines to be met through "steam stripping" rather than "air stripping." The latter simply evaporates volatile pollutants into the air, where they can pose hazards to plant workers and the public. With steam stripping, these pollutants are recaptured and potentially recycled. Recognizing that it had the authority to impose this requirement, however, EPA chose to address air pollution from this industry using its (then unspecified) Clean Air Act authority.[35] More than five years later, EPA has yet to regulate these releases under the Clean Air Act.

Second, EPA refused to require new OCPSF plants to achieve zero discharge, despite the agency's own evidence that even some existing plants could achieve this goal. In response to NRDC's challenge to this decision, a federal appeals court ruled that EPA's decision was illegal and not supported by the record, and it required EPA to reconsider.[36] Faced with this second bite at the zero discharge apple, however, EPA once again proposed to reject the zero discharge option for new chemical plants, complaining essentially that "all plants are different" and that EPA lacks the resources to evaluate this issue adequately.[37]

Standards for Indirect Discharges to POTWs

Similar problems occur due to the lack of national pretreatment standards for indirect discharges to POTWs or to the failure to update existing standards to reflect improvements in technology.

An estimated twenty-one thousand POTWs discharge wastes into U.S. waterways.[38] Sewage treatment plants are, as explained earlier, subject generally to a technology-based standard known as secondary treatment. All POTWs were to have achieved compliance with secondary treatment requirements as of July 1977.[39] While many cities did not meet this deadline, as of 1990, roughly 80 percent of POTWs were designed and constructed to provide secondary treatment or better.[40]

Secondary treatment is geared toward removal of conventional pollutants—sewage pollutants—from municipal effluent. That is an important job, but it is not enough. A large number of industrial facilities, ranging from pesticide manufacturers to small local electroplaters, discharge toxic wastes to the nation's sewage treatment plants. A 1986 EPA study identified approximately 160,000 industrial and commercial facilities discharging wastes with hazardous constituents to public sewers, representing about 12 percent of total flow to POTWs.[41]

Secondary treatment alone does not remove these pollutants. While some fraction of organic hazardous constituents is degraded incidentally in the

POTW's conventional treatment processes, many of these substances evaporate during treatment (or from the sewer pipes themselves) or wind up in the water or the sludge. None of the metals that go to a treatment plant are degraded; toxic metals pass through the plant into receiving waters or into the sludge generated by the treatment process.

Congress intended that these indirect industrial and commercial dischargers also be subject to national standards, called pretreatment standards, to control toxic pollutants they discharge to POTWs. These standards are intended to prevent exposure of plant workers to dangerous fumes or to the threat of explosion, damage to the plant infrastructure or biological processes, contamination of sewage sludge (which makes beneficial reuse more costly or impossible), and pollution of receiving waters.[42] To accomplish these goals, the act requires that the combination of pretreatment by the industrial discharger and any incidental removal by the secondary treatment process equal or exceed the level of pollutant removal achieved by BAT for a direct discharger.

Yet, as with direct dischargers, EPA's standard setting is far behind schedule. Only about one out of ten facilities discharging toxics to POTWs (14,000 to 16,000 of the more than 130,000 known) are currently subject to EPA-written national pretreatment standards.[43]

While current national pretreatment standards have made a difference in the quantities of toxics discharged to public sewers, here again we have a long way to go: *TRI* estimated that manufacturing industries discharged more than 448 million pounds of toxic contaminants to public sewers in 1990[44]—and the hazardous waste treatment industry washes another 254 million pounds per year of toxics down the drain into public sewers.[45] For the industries identified by *TRI* as reporting the largest discharges to public sewers in 1990, see Figure 5.2.

Publicly owned treatment works with total daily flow of more than 5 million gallons, and others that receive large quantities of toxic pollutants from industries or commercial facilities, are also required to have approved pretreatment programs to control the toxics discharged to their facilities. Roughly fifteen hundred POTWs are required to have approved pretreatment programs; these large plants receive an estimated 82 percent of the total industrial influent entering POTWs nationally and generate about three-quarters of U.S. sludge.[46]

Pretreatment programs are supposed to ensure that POTWs have in place the legal authority, technical capability, and resources to prevent the discharge to their facilities of toxics that could interfere with or pass through their treatment plants. In addition to enforcing the national pretreatment standards set by EPA, which apply to categories of indirect dischargers, these POTWs must have the ability to develop and enforce site-specific local limits designed to

FIGURE 5.2

Annual Discharge of Toxics into Public Sewers

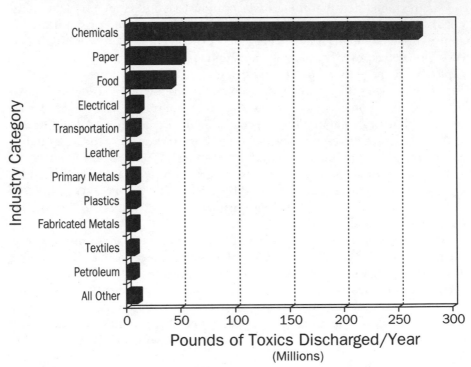

Source: EPA, 1990 Toxics Release Inventory, 58.

control additional pollutants that might pose a problem for a particular POTW (either affecting water quality, affecting sludge quality, endangering worker health, or interfering with the operations of the facility) but that are not covered by national standards.

While this requirement has been in place since 1972, more than one hundred POTWs currently violate the requirement to have an approved pretreatment program.[47] And the pretreatment programs that are in place are plagued by a wide range of problems stemming from a number of root causes.

First, since POTWs receive such a vast quantity of toxic pollutants, it is essential that they have well-written permits to control the toxics they discharge to surface waters. A POTW's permit limits dictate the strictness of the local limits that it imposes on its industrial contributors. Without a strict permit, the POTW cannot develop and implement appropriate local limits. Unfortunately, POTW permits are far from adequate. A recent EPA study concluded that

> relatively few permits issued to pretreatment POTWs contain chemical-specific limits for toxic pollutants. Most permits contain narrative restrictions...and some contain whole effluent toxicity limits.... [L]ess than 2 percent of Pretreatment POTWs are subject to limits for any single toxic organic pollutant. Limits on metals are somewhat more common, although less than 16% of Pretreatment POTWs are subject to limits on copper, which is the priority pollutant most frequently limited in NPDES permits. Since the regulatory definitions of pass through and interference...are tied to the POTW's NPDES permit limits, the general lack of toxic limits in the permits restricts the POTW's basis for developing local limits.[48]

Other inadequacies in pretreatment implementation include incorrect discharge limits for industries discharging to sewers, improper sampling of industrial effluents, improper chain of custody for samples (making it more difficult to prove and enforce against violations), undocumented inspections of industrial sites, improper sampling protocols, and inadequate enforcement against discharge violators.[49]

A 1992 EPA study of the rate of noncompliance by industries discharging to POTWs found massive violations of both reporting and discharge requirements. The study, which evaluated 1990 data on industrial dischargers to POTWs with approved pretreatment programs, found that

- 54 percent of significant industrial users of POTWs were in significant noncompliance with discharge standards and/or reporting and self-monitoring requirements;

- 35 percent were in significant violation of discharge standards (either national categorical standards or local limits or both); and

- 36 percent were in significant violation of pretreatment self-monitoring or reporting requirements.[50]

In light of these deficiencies, it should come as no surprise that POTW discharges continue to cause problems. A 1989 study by the GAO estimated that 20 percent of pretreatment POTWs had NPDES permit violations due to toxics from industries; 28 percent experienced disruption of POTW operations; and 4 percent experienced worker health and safety problems as a result of pretreatment program failures.[51] Because only a tiny fraction of POTWs even have toxics limits in their permits, from which these violations can be calculated, these statistics probably understate the extent to which POTWs are responsible for water quality problems in receiving waters as a result of industrial discharges.

Deficiencies in the pretreatment program also cause serious sludge disposal problems. The thirteen thousand to fifteen thousand POTWs in the United States generate an estimated 110 to 150 wet tons per year of sewage sludge, a by-product of sewage treatment.[52] Sludge is mostly water; the rest is organic and inorganic solids and dissolved substances. Sewage sludge contains pathogens as well as nutrients; it also may contain toxic contaminants (including heavy metals and organic compounds), depending upon what substances flowed into the POTW.

After proper water removal and odor and pathogen control, uncontaminated sludge can be a beneficial soil conditioner if handled properly and applied judiciously. A national EPA policy specifies that sludge should be used beneficially rather than become part of the waste stream clogging landfills and incinerators.[53] At present, about 36 percent of sludge is applied to forest or agricultural land, 16 percent is incinerated, 38 percent is placed in municipal landfills, and 10 percent is placed on land at sites dedicated to sludge disposal.[54]

But toxic pollutants (generally metals or organic compounds) can contaminate sludge so that it is no longer safe to use. This contamination results largely when industrial dischargers fail to adequately remove toxic pollutants from their waste streams—that is, they inadequately pretreat waste streams—before discharging to POTWs. An additional source of sludge contamination is residential and commercial wastewater discharged to POTWs; GAO estimates that residential discharges alone may account for as much as 15 percent of the regulated toxic pollutants entering POTWs[55] and that significant percentages of certain toxic pollutants are discharged into POTWs by commercial establishments not covered by the national industrial pretreatment program (such as silver from photo-processing plants and zinc, lead, and copper from car washes and radiator shops).[56]

Historically, each state has had its own regulations setting standards for beneficial reuse of sludge, particularly for land application. Inconsistencies among state programs (some of which are extremely limited in scope) and the lack of a federal bottom line establishing minimum national sludge quality standards have left gaps in public health and environmental protection.[57] Interstate conflicts also arise when states with stricter sludge quality standards seek out-of-state disposal for sludge that does not meet their own state standards.[58]

Recognizing these problems, in 1977 Congress directed EPA to set limits on the levels of toxic pollutants in sludge by 1978—limits that might, in turn, drive reductions in industrial discharges of toxics to POTWs through the pretreatment program.[59] For ten years, EPA failed to follow Congress's mandate.[60] In 1987, Congress again amended the Clean Water Act to establish another timetable for regulatory action and to specify that EPA was to set numeric limits for each sludge use or disposal practice. The agency was to issue the first set of sludge quality regulations by 1987, and its second set (for pollutants not covered in the first set due to lack of data) by 1988.[61] In addition, EPA was directed to issue rules establishing standards for state sludge management programs.[62] Sludge quality and application and disposal practices were required to be regulated under the NPDES permitting program or some other comparable permitting program that could assure compliance with the Clean Water Act.[63]

Sludge management regulations were issued in May 1989.[64] A state may apply to EPA for authority to issue sludge permits if it can show that it has in place basic legal authorities, staff resources, and enforcement ability to carry out the program.[65] In the absence of an approved state program, EPA retains control of the sludge permitting process. States have not assumed delegated program responsibility, however, because they have been waiting to see EPA's long-delayed sludge quality standards.[66]

The Environmental Protection Agency divided into several parts its efforts to write sludge quality standards. It issued one set of rules governing codisposal of sludge along with other wastes in municipal solid waste landfills in October 1991.[67] In November 1992, the administrator signed a second set of rules, governing (1) land application of sludge, (2) distribution and marketing of sludge products to the public, (3) surface disposal sites, and (4) sludge incineration.[68]

Each of EPA's sludge quality rules has been highly controversial. First, in apparent violation of the statutory mandate,[69] the municipal landfill (codisposal) regulations set no numeric limits for sludge quality; under EPA's rule, the sludge going to municipal landfills can be highly contaminated so long as it does not qualify as a hazardous waste under RCRA.

This rule also allows POTWs to issue "removal credits" for an unlimited list of toxic pollutants. These credits allow an industrial user to increase the amount of toxic pollutants it discharges to a POTW[70] on the theory that landfill restrictions will prevent the toxic-laden sludge from leaching out of the landfill. This arrangement creates a perverse incentive for industries sending wastes to POTWs, particularly those that release metals and other toxics that concentrate more readily in sludge: If an industry can persuade its POTW to place its sludge in a landfill, it can seek removal credits to *increase* the toxics it emits. This outcome runs directly counter to the national policy to support the beneficial reuse of sludge.[71] Because of these and other problems, NRDC and the Sierra Club challenged the codisposal rules in court.[72]

The regulations for other sludge disposal and sludge use methods were signed in late 1992 following a storm of controversy.[73] The law requires that each sludge generator be in compliance with applicable standards no later than twelve months after promulgation (or two years if new pollution control facilities are needed).[74] It also requires that sludge limits be implemented through NPDES permits or an equivalent permitting scheme.[75] Despite these requirements, EPA has decided that it will not revise (or require states to revise) NPDES permit limits to cover sludge until the next time the permits are due for renewal—although sludge generators are expected to comply with the rule's reporting, record-keeping, and monitoring rules immediately.[76] This self-implementing system, intended to save state and federal resources, raises many red flags. A 1990 GAO report raised a number of serious concerns about the efficacy of the sludge regulatory scheme EPA had devised—and this was before EPA weakened the scheme still further by delaying sludge permitting requirements.[77]

The new rules for land application, incineration, land disposal, and distribution and marketing establish numerical criteria for toxic contaminants in each disposal and use practice: ten metals for land application and distribution and marketing, three metals for land disposal sites, and six metals, as well as "total" organic compounds, for incineration.[78] The debate over whether the rules provide adequate public health and environmental protection is likely to continue, since EPA is charged with reviewing and revising the rules every two years.[79]

Permit Program Deficiencies

Technology-based and water-quality–based standards are implemented through EPA and state permitting and enforcement programs. Unfortunately, a large number of sources remain unpermitted or subject to unduly weak permit requirements. Permits for polluted urban storm water and raw sewage from CSOs are particularly lacking. And many permit requirements that do exist are enforced weakly or, in many cases, not at all.

Even the most protective water quality standards and the strictest technology-based standards must be translated into enforceable pollution controls through permits written by EPA or state officials.

Because states and EPA regional offices run the NPDES program, and because it is implemented through thousands of individual permits, less empirical information is available on the adequacy of the permitting system than on the national regulatory program. It would be impossible, at least without an army of auditors, to review each permit to evaluate the quality of NPDES permits around the country. Instead, the quality of the program must be deduced from available information about individual states and regions and, in some cases, anecdotal information about specific permits.

Unfortunately, here too we find serious problems with the way in which the Clean Water Act is being implemented in the real world. Many sources lack permits altogether. Existing permits are often weak, incomplete, and outdated. While it is impossible to identify all of the deficiencies in the existing permitting system, some of the most glaring problems are outlined on the following pages.

Unpermitted Discharges

EPA estimates that ten thousand dischargers have no permits whatsoever, in violation of the clear requirement of the Clean Water Act.[80] Several studies suggest that unpermitted facilities may be a significant source of pollution in at least some areas. In Puget Sound, Washington, for example, where many non-permitted commercial and industrial facilities were identified, as much as 20 percent of the toxics entering Puget Sound were estimated to originate from such nonpermitted discharges.[81] And astonishingly, EPA's own inspector general found that "[r]egion 6 has failed to issue any [NPDES] permits for oil and gas discharges into Louisiana wetlands," despite the fact that "[s]cientists, environmentalists, and even other Federal agency officials believe that current petroleum industry discharges are causing serious degradation of water, wetlands, and associated fish and wildlife resources throughout coastal Louisiana."[82]

The largest categories of unpermitted or inadequately permitted point sources are for municipal and industrial storm water and CSOs. These types of discharge result in massive—and essentially unregulated—releases of pollutants into the nation's waters.

Urban and Industrial Storm Water

The Environmental Protection Agency's Nationwide Urban Runoff Program (NURP) report has provided the primary evidence on the ubiquitous nature and significant quantities of pollutants in urban runoff: "Data collected under the NURP indicated that on an annual loading basis, suspended solids in discharges from separate storm sewers...are around an order of magnitude greater than solids in discharges from municipal secondary sewage treatment plants."[83]

Chemical oxygen demand, fecal coliforms, and oil and grease in urban runoff are also significant. Urban runoff, according to EPA, is "an extremely important source of oil pollution to receiving waters."[84] Construction runoff is also a massive problem:

> Intensive construction activities may result in severe localized impacts on water quality because of high unit loads of pollutants, primarily sediments.... Sediment loadings rates from construction sites are typically 10 to 20 times that of agricultural lands, with runoff rates as high as 100 times that of agricultural lands, and typically 1,000 to 2,000 times that of forest lands. Even a small amount of construction may have a significant negative impact on water quality in localized areas.[85]

Studies conducted by NRDC in U.S. cities, including Washington, D.C., have shown that the contribution of urban runoff to total annual pollutant loadings to urban streams and estuaries rivals, and in some cases surpasses, loadings of the same pollutants from factories and sewage plants.[86]

Storm water also contains toxic chemicals: 77 of 120 toxic priority pollutants monitored were found in storm water from residential, commercial, and light-industrial lands.[87] Of these toxics, 24 were found in more than 10 percent of all samples taken.[88] The Environmental Protection Agency's compilation of industrial stormwater data from the Permit Compliance System shows that industrial stormwater discharges, though variable in quality and quantity, are an important, often uncontrolled source of water pollution. For example, storm water from chemical manufacturers is contaminated with cyanide, phenolics, and trichloroethylene.[89]

Finally, the flow of storm water itself damages streams and significantly degrades water bodies in urban areas. The severe physical habitat effects that can result from stormwater discharges can include streambank erosion and rapid changes in stream channel morphology, loss of protective riparian trees and other vegetative cover, and loss of pool and riffle structures. All of these changes, caused by high-volume and high-velocity storm flows that occur frequently after the construction of impervious surfaces such as parking lots, destroy the physical habitat necessary to support fish and other aquatic life.

Despite these severe problems, EPA has taken almost twenty years, with constant prodding from environmental groups, Congress, and the courts, even to begin to deal with storm water. In 1973, EPA acknowledged that stormwater discharges fall within the Clean Water Act definition of point source. However,

EPA regulations exempted many stormwater discharges from NPDES permit requirements, arguing that there were simply too many outfalls to regulate.[90]

The Natural Resources Defense Council successfully challenged EPA's authority to exempt these and other point sources from regulation.[91] This "successful" court challenge, however, was a classic case of winning the first battle in what would be a very long war. Over the course of the next decade, EPA issued and reissued stormwater control regulations, only to have them challenged in court by industries and cities as too strong, or challenged by NRDC as too weak, or withdrawn by the agency itself as administrations and policies changed. Separate proposed or final rule-making notices were issued in 1979, 1980, 1982, 1984, and 1985.[92]

In the 1987 Clean Water Act Amendments, Congress recognized the role of contaminated stormwater runoff in the severe, ongoing pollution of this nation's waters and expressed impatience with EPA's slow progress: "[S]tormwater runoff from urban areas contains large volumes of toxic materials and other pollutants. Since 1972, municipal separate storm sewers have been subject to the point source permit requirements of the Clean Water Act. However, EPA only recently began to develop a permit program for these sources."[93] To address this long history of delay, in 1987 Congress put EPA on a new, phased schedule for regulating storm water. EPA was to regulate first large cities (systems serving more than 250,000 people) and industries, then medium municipalities (systems serving more than 100,000 people), and finally small municipalities and other sources.[94]

Once again, EPA missed the law's deadlines for issuing stormwater regulations. The first phase of stormwater rules was issued nearly two years after the deadline (November 16, 1990). Again, EPA included major loopholes in its rules—for example, exemptions for a large number of industries and construction sites, which were once again rejected by a federal court after a legal challenge by NRDC. And EPA still has not issued rules for small cities and other sources of storm water.[95]

Two decades after the 1972 law was passed, where do we stand on controlling the massive releases of pollutants from stormwater outfalls? The vast majority of industrial sources of storm water are just now filing permit applications, as are municipal systems serving more than 100,000 people. Most permits have not actually been issued, and compliance with these permits is years off. The Environmental Protection Agency is revisiting the loopholes rejected by the court for construction sites and so-called light industries, but it announced (perhaps illegally) that these sources will not require permits in the interim. Faced with EPA's continued intransigence on other sources of storm water, Congress gave the agency yet another year to issue the remaining rules, and it gave these sources another two years to file permit applications. It appears that the new century will arrive before all sources of storm water have even the most basic permits under the Clean Water Act.

Combined Sewer Overflows

Combined sewer overflows (CSOs) are discharges of raw (untreated) human sewage mixed with stormwater runoff and, often, industrial wastes; they occur during storms when older sewer pipes carrying both storm water and raw sewage become overloaded. During overflows, the foul mixture is diverted from the sewage treatment plant at overflow points and is discharged directly into our rivers, lakes, and coastal waters.

The CSO problem is a big one. According to EPA estimates, more than 1,100 collection systems in the United States, serving approximately 40 million people, have combined sewer systems.[96] Of those, 328 systems, serving about 25 million people, had documented needs for wastewater treatment control as of 1988.[97] The documented price tag to fix these problems totaled $16.4 billion in 1988 dollars.[98] More than half of these CSO needs are located in marine and estuarine systems that serve approximately 12 million people.[99] Thus, while the number of systems around the country that have CSOs is relatively small in comparison to the total number of sewage treatment systems (approximately 1,100 out of more than 24,000 collection and treatment systems),[100] the affected systems serve nearly 16 percent of the nation's population.[101]

A number of large U.S. cities have significant CSO problems. A 1992 NRDC study showed that fourteen large cities with CSOs discharge more than 165 billion gallons of raw sewage mixed with polluted stormwater and industrial discharges (see Table 5.2).

The quantity and range of pollutants discharged by CSOs are significant. While precise figures are not available due to a shortage of monitoring data, the following estimates can be made:

● Total suspended solids in CSO discharges can range from roughly 400 to 700 milligrams per liter (mg/L)—roughly two to three times the suspended solids concentration of "normal" raw sewage.[102]
● Biological oxygen demand (a measure of the organic material in water that can rob aquatic life of oxygen) ranges in concentration in CSOs from roughly 80 mg/L to 150 mg/L.[103] (In comparison, the effluent from a sewage treatment plant is required to meet a thirty-day average discharge standard of 30 mg/L.[104])
● Fecal coliform counts in CSO discharges can range from two hundred thousand to more than 1 million per 100 milliliters.[105] (In comparison, many

TABLE 5.2

Estimated Annual Combined Sewer Overflow Releases from Fourteen U.S. Cities (In Pounds)

CITY	GALLONS (BILLIONS)	ORGANIC SEDIMENTS (MILLIONS)	WASTES (MILLIONS)	COPPER	LEAD	ZINC
Atlanta	5.3	1.5	5.5	4,500	15,000	15,000
Boston	5.2	9.4	4.3	3,900	7,900	11,000
Bridgeport	1.7	1.1	0.4	1,500	4,900	5,000
Chicago	27.0	10.0	6.9	21,000	4,400	144,000
Cleveland	5.9	26.0	4.7	6,700	9,200	12,000
Minneapolis-St. Paul	1.6	2.5	0.7	800	1,500	3,500
Narragansett	2.6	3.5	2.3	3,000	1,700	7,000
New Bedford	1.1	1.7	0.5	1,000	3,300	3,300
New York City	84.0	83.0	38.0	71,000	240,000	240,000
Philadelphia	20.0	23.0	17.0	17,000	58,000	58,000
Richmond	4.1	9.1	2.5	3,500	12,000	12,000
San Francisco	1.7	1.8	1.5	1,500	4,900	5,000
Seattle	2.9	2.8	1.5	2,100	4,200	5,400
Washington, D.C.	2.2	5.4	0.9	1,900	5,500	5,200
TOTAL	165.3	194.3	86.7	223,200	372,500	525,400

Source: NRDC, *When It Rains...It Pollutes* (April 1992).

states recommend that bathing be restricted at beaches where fecal coliforms exceed two hundred per 100 milliliters.[106])

For the estimated total annual national mass loadings from CSOs for selected pollutants, see Table 5.3.

Little information is available on the toxic pollutants discharged by CSOs. However, since 12 percent of the total flow to sewage treatment plants nationwide consists of industrial wastewater,[108] industrial toxics are likely to be present in significant quantities in CSO discharges; industrial flow to these systems receives no treatment at the POTW during overflow events. And, as previously discussed, urban stormwater runoff, which makes up the bulk of total volumes of CSO discharges, also contains numerous toxic pollutants, including heavy metals, oil and grease, and organic chemicals.

While major progress has been made in the past decade in achieving a minimum level of secondary treatment for municipal POTW discharges, the CSO control program has lagged far behind. One principal reason is that, while CSOs have been eligible for some portion of construction grant financing, relatively few federal funds actually have been available for CSO abatement. The magnitude of CSOs' contribution to water quality problems also may have been obscured to some degree by sewage treatment plant discharges until the secondary treatment program was well on its way.

Until recently, EPA did little on a national level to bring CSO discharges under control, either because it did not wish to press the states to take action when it lacked a major federal grants program to assist them or because it had other priorities. Although EPA maintains that CSOs are subject to BAT standards, it has not drafted a national categorical standard for CSO discharges but instead has left state or regional permit-writers no alternative but to rely upon their BPJ for CSO permitting.

TABLE 5.3

Estimated Annual CSO Pollutant Discharges		
POLLUTANT	RANGE OF LOADING FACTOR (LB/ACRE/YR)	RANGE OF NATIONAL CSO LOADING (LB/YR)
Total solids	(TDS & TSS)*	1,986 to 4,342 2.7 to 11 billion
BOD*	446 to 1,188	1.1 to 3 billion
Ammonia	46 to 75	120 to 188 million

Sources: Derived from J. Bryan Ellis, "Pollutional Aspects of Urban Runoff," at 20, reprinted in *Urban Runoff Quality*, H.C. Torno, J. Marsalek, and M. Desbordes, eds. (Springer, Verlag, Heidelberg, 1986), 20.[107]
* TDS = total dissolved solids; TSS = total suspended solids; BOD = biological oxygen demand (oxygen-robbing matter).

Elusive Zero Discharge Goal

In August 1989, EPA did publish a national CSO control strategy. The strategy called upon all states (or EPA, where it is the permit-issuing authority) to develop and submit to the agency, by January 15, 1990, a statewide permitting strategy for CSO controls. The permitting strategy required that the responsible entity do the following:

1. Identify communities with CSOs, including each CSO point, and determine whether the CSOs are subject to permits and, if so, whether they are in compliance with applicable standards;

2. Set priorities for achieving compliance and describe how compliance will be achieved (including descriptions of the nature of control measures to be applied);

3. Issue permits (where possible, on a system-wide basis rather than for individual outfalls) for each CSO system. Permits are to include monitoring requirements and permit reopener clauses based on results of the testing; and

4. Establish compliance schedules in those instances where statutory deadlines cannot be met, using administrative enforcement orders or other legal enforcement tools.

The CSO strategy also gave extremely limited guidance as to what constitutes BAT, the minimum level of technology required under EPA's interpretation of the law: proper operations and maintenance, maximum use of the sewer collection system for storage, pretreatment program revisions to minimize CSO impacts, maximization of the POTW's capacity to accept storm flows, prohibitions on dry-weather overflows, and control of solids and floatables. The strategy also notes that, where water quality standards are not met, additional controls must be placed on discharges. But it still leaves to individual permit-writers the job of deciding what will actually be required to control each CSO permittee.

According to EPA, thirty states and regions had submitted strategies as of July 20, 1992, all of which have been approved; another five states have combined sewers but claim that they need no strategy.[109] Meanwhile, recognizing the inadequacy of the existing permitting strategy, EPA has committed to issuance of a revised policy.

During the summer and fall of 1992, EPA convened a policy dialogue in which representatives of cities, states, and environmental groups attempted to reach consensus on a new CSO permitting strategy. While the parties did not reach full consensus, a joint framework for a new strategy ultimately was presented to EPA. Building on this framework, in late December, EPA released for public comment a new draft CSO permitting strategy.[110] In this proposal, EPA added the following to the six earlier minimum controls: pollution prevention (including water conservation), public notice of waters affected by CSO discharges, and adequate monitoring. In a major step forward, the new strategy also proposes a menu of minimum technology-based controls from which cities may choose.[111] In addition, the proposal would require relocation or elimination of releases currently going into sensitive waters wherever feasible.

Backlogged Permits and Antibacksliding

The Clean Water Act requires that NPDES permits be issued for no more than five years. The review and renewal of NPDES permits presents a critical opportunity to include new limits based on updated effluent guidelines or on more comprehensive water quality standards. The five-year permit term also ensures a regular opportunity for public comment and participation.

Permit backlogs—permits that have expired—have plagued the NPDES program since its inception. Under EPA rules, federally issued permits are extended automatically, no matter how out-of-date their limits, if the permittee has submitted a renewal application. Many states operate the same way. This means that the facilities continue to discharge with impunity, often for years, after the expiration of their five-year permits.

Unfortunately, this problem is getting worse rather than better. According to recent EPA estimates, "The number of backlogged permits in the States has increased from 12% in 1988 to 22% in 1990.... The result of this increased backlog is that some permits needing new effluent limitations to protect water quality are not being reissued."[112]

A related issue is the degree to which permits can be weakened when they are renewed or reissued. Such actions are particularly troublesome because they move away from, rather than closer to, the zero discharge goal of the Clean Water Act. In 1987, Congress established an "antibacksliding" provision to limit the weakening of permits to narrowly prescribed circumstances, such as when mistakes were clearly made during issuance of the original permit or when plant expansion necessitates higher discharge limits.[113] Disturbingly, however, five years after this provision was added to the law, EPA still has not issued implementing regulations. While the provision itself is self-executing, absent federal rules, it is unclear to what extent EPA and state permit-writers have honored the requirement.

Water-Quality–Based Permits

Where technology-based requirements (for industry) or secondary treatment requirements (for POTWs) are not enough to protect water quality in a particular location, the Clean Water Act requires that NPDES permits be made more stringent. These "water-quality–based" permits are supposed to ensure that a discharge does not cause water quality standards violations in the receiving lake, river, or estuary.

As previously discussed, there are serious problems with the numerical criteria that states have adopted (or failed to adopt) to protect receiving waters. And because of the serious gaps in our monitoring and permitting efforts, we do not know the full scope of waters in which quality standards are violated. The Chesapeake Bay Foundation, for example, found in a survey of dischargers in 1989 that few POTWs had necessary water-quality–based effluent limits for nutrients and toxics.[114] Similarly, a sampling survey in two EPA regions showed that more than one-third of the facilities currently discharging exhibit toxicity at levels having a reasonable potential to violate at least one water quality standard.[115]

But the problems with translating water quality criteria into meaningful permit limits go far deeper. Due to gaps and loopholes in EPA guidance and regulations, even where water-quality–based permits are issued, often they will not protect receiving waters adequately.

At the heart of the matter is the questionable concept of dilution. The framers of the act were clear that water quality goals be achieved through pollution reduction rather than dilution: "The Conference agreement specifically bans pollution dilution as an alternative to waste treatment."[116] But compliance with this intent is far from clear, leading two commenters to denounce the "dilution of the Clean Water Act."[117] While the act clearly requires that water quality standards be met in the receiving water, considerable controversy has surrounded the related questions of how end-of-pipe limits should be calculated to achieve this goal, and of where the standard must be met.

EPA and states have grappled with questions of how to predict the impact of a particular discharge on a specific water body and of what assumptions should go into that equation. Upstream levels of a pollutant affect how much a discharger may add before it is predicted that water quality standards will be violated. The amount of water available to "assimilate" these wastes also dictates end-of-pipe requirements.

The law instructs that these judgments be made with "a margin of safety which takes into account any lack of knowledge concerning the relationship between effluent limitations and water quality."[118] However, too often these questions are answered in favor of the polluter rather than the river or lake, which rarely has a representative at the table when permit limits are being negotiated. These permits are supposed to be written based on a detailed total maximum daily load (TMDL) and wasteload allocation (WLA), which take into account information about all sources of a pollutant into a given receiving water.[119] The TMDL calculates the total amount of a given pollutant that can be released into a water body before the water quality standard will be violated, and it depends on a host of variables and assumptions about the fate and effect of specific pollutants. The WLA allocates this total amount of permissible discharges among all dischargers of that pollutant to a particular waterway.[120] Permit-writers are supposed to assure that multiple sources of pollutants (point, nonpoint, and background) will not gang up on the unsuspecting waterway. But TMDLs and WLAs have not been calculated for many waters that require water-quality–based limits. A 1989 survey of four EPA regions by GAO found that this analysis was performed for only a small fraction of polluted waters:

> Many states and EPA have not developed total maximum daily loads for many of the nation's most polluted waters. Information available at and discussions with officials of EPA Region X [Seattle] showed that maximum loads had been set for 1 of the 602 water-quality-limited segments. Maximum loads are now being developed or planned for 41 of the remaining 601 water-quality-limited segments. Similarly, maximum loads had been set for only 4 of EPA's New York Region's 168 segments. More maximum loads had been set in the other two regions GAO contacted: 43 percent had been set for EPA's San Francisco Region's 77 segments, and 31 percent had been set for Chicago Region's 227 segments during fiscal year 1987. *Consequently, 16 years after the Clean Water Act was enacted, many water segments did not meet state standard* [121] (emphasis added).

In an effort to jump-start the process of writing water-quality–based permits, in 1987 Congress added a new "toxic hotspots" provision to the Clean Water Act (section 304(1)). Under this provision, states were required to identify all water bodies that violate water quality standards or designated uses due to toxic or other pollutants, identify point source dischargers that contribute to those problems, and write new permit limits to control those releases.[122] There are several problems, however, with the implementation of this program. First, based on NRDC's review of the efforts of two states to identify polluted waters, many more such waters exist than were listed under the program.[123] Moreover, as reviewed extensively in chapter 2, state views on which waters are polluted ignore many impacts to human health and the environment.

Nevertheless, the state listing process under section 304(l) identified more than seventeen thousand waters around the country that fail to meet water quality standards or otherwise do not support designated uses. Despite these listings, under a narrow EPA interpretation of the law, states were required to impose new control requirements for only a fraction of these waters (less than seven hundred of the seventeen thousand). This view was challenged by NRDC and rejected by the Ninth Circuit Court of Appeals,[124] but on review and reconsideration, EPA has proposed not to expand the program.[125]

Second, even where TMDLs and WLAs are written and water-quality–based permits are issued, serious questions remain about whether the standard must be met at all places within the water body or whether limited violations may occur in the immediate vicinity of the outfall, to allow the effluent to "mix" with the receiving water. Professor Houck referred to this as "the wildest card of all: mixing zones."[126] In recent years, this situation has become even more bizarre, with many states adopting a "mixing zone within a mixing zone," known as zones of initial dilution, or ZIDs. Under this practice, acute water quality impacts are permitted within the smaller ZID, while chronic impacts may occur within the larger mixing zone.

Environmental Protection Agency regulations are virtually silent with respect to the manner in which states can calculate water-quality–based permits: "States may, at their discretion, include in their State standards, policies generally affecting their application and implementation, such as mixing zones, low flows and variances. Such policies are subject to EPA review and approval."[127]

The agency has issued considerable guidance on how states *should* implement these policies. But such guidance is not, by itself, legally binding.

Worse, EPA's guidance increasingly gives states wider latitude to use dilution as an alternative to treatment, and in many cases it actively encourages such practices. The slippery slope is revealed by the fact that EPA's 1983 general guidance on mixing zones advises that acute toxicity should not be allowed *anywhere* within a mixing zone;[128] yet EPA now accepts the use of ZIDs that allow acute toxicity within the smaller zone of dilution.[129]

Proponents of these procedures argue that requiring water quality standards to be met at the end of the pipe is "treatment for treatment's sake," with little or no actual benefit to the receiving water. This argument trivializes the zero discharge philosophy of the Clean Water Act, as well as the goal of no toxics in toxic amounts. More important, it ignores the fact that many pollutants are persistent, that is, remain in the aquatic ecosystem (including aquatic sediments) for long periods of time; and others are bioaccumulative, building up over time in fish, birds, mammals, and other species. To guard against these effects, we need to consider the total mass of toxic pollutants released, not the concentration of pollutants in the water column after giving every conceivable allowance for dilution.

Mixing zones are just one way of using dilution and other factors to change the stringency with which water quality criteria are translated into enforceable permit limits. While some of these issues are technically complex and beyond the scope of this book, Dr. Foran identified the following major examples:[130]

- Design flows. States use different assumptions about how much receiving water is available to dilute the concentration of pollutants outside of the mixing zone.

- Detection limits. In some cases, the concentration of pollutant in an end-of-pipe effluent limit necessary to meet the water quality criterion is below the detection ability of existing instruments. Different ways of defining and implementing detection limits can affect the stringency of the permit.

- Background concentrations. The amount of pollutants a discharger may release without causing a water quality standard violation may depend on the presence of the same pollutant in the receiving water or intake water. Different ways of handling this issue result in different permit limits. In some cases, state allowance for background concentrations guarantees water quality violations.

- Multiple pollutants. Pollutant-specific water quality criteria do not account for the cumulative or synergistic effects of multiple pollutants. Some states do not account for this at all; others do so in varying ways. To address aquatic life effects, some states use whole effluent toxicity (WET) testing, which measures the direct toxicity of the effluent itself to a range of test species. However, great variation exists in the methods of application and in the resulting stringency of permits. Moreover, WET testing cannot measure long-range impacts in the water body itself. There is no similar accepted methodology for the human health effects of multiple pollutants. Such impacts can be addressed by lowering the allowable release of each pollutant, but permit-writers do not generally do this.

- Fate and availability. Many toxic pollutants do not remain dissolved or suspended in the water column. Some bind to particles and can contaminate the sediment below. Some states argue that different pollutants are more or less biologically available, that is, in forms that do not affect fish or aquatic life. These states may write weaker permit limits based only on the estimated available percentage of the pollutant. Such corrections, however, do not account for factors such as subsequent resuspension of particles and uptake by organisms that live in or feed from the sediment.

Another major, troublesome issue lurking in the background of NPDES permitting for both storm water and CSOs is the controversy over so-called wet-weather water quality standards. Arguing that water quality standards are designed for low-flow conditions, some cities propose that revised (presumably weaker) standards be set for high-flow conditions. These arguments misunderstand the fact that water quality standards themselves are established independent of flow and are only applied in permits to ensure that in-stream standards will not be violated under most (therefore low-flow) conditions. During storm flows, if anything, much higher total amounts (mass) of pollutants are released, justifying stricter rather than weaker standards.

The Environmental Protection Agency gives guidance but has no binding regulations to establish national consistency on these issues. Thus, states are given virtually a free hand to devise loopholes in the system of water-quality–based permitting. Aside from the fact that this results in continuing discharges of toxics in toxic amounts, it leads to serious variations in the degree of protection that citizens and aquatic ecosystems receive in different, sometimes even neighboring, states. A recent study of water-quality–based permits, commissioned by the IJC, identified a number of reasons for inconsistent and unprotective permits: "A series of what can only be termed compromises has been incorporated into the regulatory programs of U.S. jurisdictions; these compromises are designed to regulate the discharges of persistent toxic pollutants. *The compromises may result in the discharge of very large loads of persistent toxic pollutants into the Great Lakes ecosystem*"[131] (emphasis added). The study found that:

● Substantial differences exist in the applicable numerical water quality criteria for the toxic pollutants studied. There was a range spanning three orders of magnitude for mercury, for example, depending on the jurisdiction. New York would allow the hypothetical permittee to discharge more than 145 kilograms per year (kg/yr) of mercury, while Michigan would allow the same entity to release less than 1 kg/yr. Similarly, Illinois would permit the discharger to release 7,000 kg/yr of lead, while Michigan would permit only 700 kg/yr.[132]

● The results of differences in the use of mixing zones, dilution, and other elements of state programs yielded "highly variable limitations on the concentrations and loads of toxic pollutants that are discharged from point sources" in the different jurisdictions.

● *Large loads were discharged by [a] hypothetical industry in each and every state, although the loads were highly variable.* The range between the highest and the lowest loads of pollutants discharged by the hypothetical industry, using each state's regulatory procedures, spans an order of magnitude or more.

● Even in states that appear to have stringent numerical water quality criteria for toxic pollutants, permissive assumptions about dilution flows offset those stringent criteria.[133]

In essence, the Great Lakes study confirms what many citizens who have challenged individual permits have learned the hard way: There is more to effective control of toxic pollutants than a good numerical criterion for the pollutant. At many steps along the way, by failing to police more rigorously the manipulation of flows, concepts of mixing, and assumptions about background levels of toxics already present in the receiving waters, EPA has allowed states to continue to foul their waters with toxics under the guise of water-quality–based analysis.

Technology-Based Permits

In theory, translating technology-based requirements to end-of-pipe effluent limits should be subject to less manipulation than for water-quality–based permits. After all, Congress intended technology-based requirements to "level the playing field" among similar dischargers, requiring uniform pollution reductions regardless of location or receiving waters.[134]

Since so many industries are not subject to categorical effluent limitations or pretreatment standards, but instead are subject to case-by-case BPJ permits, however, dischargers are subject to vastly different control obligations. A 1984 EPA study found that state permit-writers had considerable difficulty in writing defensible BPJ permit limits, due to a range of problems, including inadequate training and lack of access to information and resources.[135]

One of the inadequacies in BPJ permitting is apparent from EPA's recent proposed effluent guidelines for the pesticide-manufacturing industry. Astonishingly, the preamble to the proposed rule revealed that twenty-one existing pesticide manufacturers did not know what priority toxic pollutants were in their discharge.[136] Apparently, in two decades, the BPJ permit-writers for these facilities had not required the most elementary first step of the permitting process for toxic pollutants—monitoring to see what is in the discharge. Clearly, the permits did not regulate any toxic pollutants for these facilities, in plain violation of the law and EPA rules.[137]

Even the existence of national guidelines does not assure adequate technology-based permit requirements. This issue has not been the subject of serious regional or nationwide studies. To determine whether technology-based permits are applied in a relatively uniform fashion would require a massive examination of a large number of individual permits. But anecdotal evidence indicates that the permitting system can be (and is) "gamed" to achieve weaker treatment and discharge requirements than are warranted. (The NPDES program and its regulations are extremely complex; we present here only the most obvious ways in which actual permit limits can be affected. Many other examples exist as well, such as the opportunity to take "credit" for pollutants in intake water.[138])

First, some effluent guidelines are written in terms of the mass of pollutants that may be discharged per unit of production, for example, x kilograms of cyanide per metric ton of product produced.[139] A permit-writer translates this general rule by determining how much the plant will produce over the permit term (five years) and then calculating the end-of-pipe limits. (The process may be somewhat more complex, for example, where effluent limitations are expressed as both instantaneous and daily limits.) Thus, the choice of how to determine production can alter seriously the amount of pollution the plant may release. Should this be based on historical or projected production? Average annual or highest annual production? How much effort can overworked permit-writers put into verifying production data provided as part of the permit application? The Environmental Protection Agency's regulations provide little firm guidance on these issues,[140] leaving actual limits open to substantial manipulation.

A review of an NPDES permit application for an aluminum-manufacturing facility revealed just such "gamesmanship." By comparing the production figures submitted in the permit application with actual production levels reported in a company periodical, the reviewer found that the applicant seriously overestimated production in two ways. First, the company used maximum rather than average daily production rates; second, the company used old production data, despite the fact that production levels had been reduced dramatically.[141] A similar case was revealed when NRDC and the Southern Environmental Law Center commented on a permit application for a pulp-and-paper plant in North Carolina. Like the applicant for the aluminum plant, the pulp-and-paper plant applicant apparently inflated production levels in an attempt to gain higher permit limits.[142]

Second, some effluent guidelines are expressed in terms of concentrations of pollutants in the discharge—for example, x micrograms of benzene per liter of effluent released.[143] Again, while this system appears relatively straightforward, it is subject to gamesmanship. For example, a given pollutant concentration at the end of the pipe can be achieved by adequate treatment, or it can be achieved by diluting a certain process waste with another waste stream in which the same pollutant is not present, or with nonprocess waters, such as cooling water or stormwater runoff. While EPA could address the problem of dilution by requiring monitoring at the end of each waste stream instead of at the end of the pipe, its regulations not only make this practice optional, they actively discourage it.[144]

Incomplete and Inadequate State Programs

States can assume responsibility for their NPDES permitting programs from EPA. In order to be delegated this authority, each state must demonstrate that it has the legal authority, staff, and technical capability to manage the program.[145] Obviously, even the most stringent federal program, which is hardly in place, is meaningless without sound state implementation programs. The state also must agree to assume responsibility for the pretreatment program[146]—it would be administratively cumbersome for a state to issue the NPDES permit for a POTW while EPA retained control over the program regulating industrial discharges to those POTWs.

To date, thirty-nine states have been delegated authority to manage their NPDES programs. Yet, of those thirty-nine, twelve still do not have authority to manage their pretreatment programs.[147] This may be one reason why both the pretreatment program and the NPDES program continue to experience so many problems.

We are aware of no effort similar to EPA's comprehensive review of the pretreatment program, which Congress mandated, to review the adequacy of state NPDES programs around the country. But problems with these programs abound, as evidenced by the fact that environmental groups have petitioned EPA to revoke NPDES program delegation or formally expressed concerns about program inadequacies in six states since 1989. The Environmental Protection Agency itself has take action to address serious problems in another two states. Issues range from legal deficiencies in citizen access to the courts to pervasive defects in NPDES pretreatment and enforcement programs.[148]

Enforcement Deficiencies

Even assuming that good permits are in place for industrial and municipal dischargers, effective progress is made only if these permits are obeyed. Sad to say, inadequate enforcement is another major problem with both the NPDES and pretreatment programs. Study after study has documented that dischargers, both direct and indirect, violate the law repeatedly and flagrantly—and get away with it nearly all the time.

A 1991 national study conducted by the U.S. and New Jersey Public Interest Research Groups (PIRGs) assessed compliance by major facilities around the country and found extremely high levels of noncompliance with permit limits and reporting requirements. PIRG found that

[i]n just the three month period from July through September of 1990, which represent[ed] the most recent data available from EPA [at the

time of the study], 12% of the largest industrial facilities in the nation and 13% of the largest municipal facilities were...in Significant Noncompliance with the Clean Water Act. Fifteen percent of the nation's Federal facilities were...in Significant Noncompliance with the Act. Overall, 49 States and all 7 Territories reported major industrial, municipal and/or federal facilities in significant violation.[149]

More than half of the industrial and one-quarter of the municipal violators were reported as significant because of their failures to submit discharge monitoring reports. "The frequency of under-reporting of data indicates that this may not always be due to an oversight on the part of violators, but instead, may be due to a deliberate attempt to mask serious violations."[150]

Regional analyses of noncompliance yield similarly dismal results. For example, an EPA review of federal facility noncompliance in the Chesapeake Bay region during 1989 and 1990 uncovered 50 of 311 federal installations with significant violations of environmental laws, including discharge permits.[151] And the Chesapeake Bay Foundation's review of POTWs in the Bay region found that only 23 percent were in full compliance, and 36 percent violated permit limits during four or more months in 1989.[152]

In addition to reported instances of noncompliance, many facilities fail to meet Clean Water Act requirements, masked through the fiction of their placement on "schedules of compliance" through the enforcement process. Under these schedules, a violator is given less stringent enforcement limits for a period of time while improvements are made to bring the facility into compliance with actual permit limits. In some cases, compliance schedules even extend beyond the permit term. Since these facilities are meeting their less stringent requirements, they do not appear in EPA's reporting system as being out of compliance, even though they are not meeting actual permit limits. If these facilities meeting less stringent permit limits during the study period were added to the count of facilities in significant noncompliance, the level of measured noncompliance would jump dramatically to 20 percent for all major facilities, including 22 percent for municipal facilities, 17 percent for industrial facilities, and 25 percent for federal facilities.[153]

The Clean Water Act's toxic hotspot program, discussed previously, mirrors this illusion of compliance. A GAO review of 529 facilities on the toxic hotspot list revealed that, in many cases, the permits issued to these facilities did not actually contain any more stringent permit limits; they were simply modified with the addition of a three-year compliance deadline. Some of the dischargers targeted by the toxic hotspot program opted to hook up to sewage treatment plants, GAO found, becoming indirect dischargers without actually controlling their releases of toxics.[154] In light of the poor compliance rates for indirect dischargers previously discussed, this hardly gives reason to anticipate that better toxics control will result.

Another significant reason for the inadequacy of existing enforcement is that penalties, when they are assessed, are too low to offer meaningful incentive to comply. Civil penalty rates do not deter violations or protect water quality because they do not recapture the economic benefits of noncompliance; that is, it may be cheaper to pay the penalty than to comply with the law.

The EPA inspector general issued a draft report in 1989 analyzing the agency's application of its civil penalty policies under its various programs, including water, and stated:

> Appropriate penalties were either not calculated and assessed at all, or inadequately calculated. Also, calculated penalties were reduced during negotiations, in some cases in excess of 90% and amounting to millions of dollars, with little or no documentation to support the reductions. In many cases the financial benefits the violator received from delayed or avoided costs were not recovered.[155]

The situation is not much better when one moves to criminal court, the presumed remedy for the most flagrant or persistent polluters (see Table 5.4). An NRDC review of criminal enforcement under the Clean Water Act over six years[156] showed that, despite some recent progress, punishment for criminal violations of the act remains low. For individuals, monetary fines have increased in recent years but remain just above the statutory minimum and about one-fifth of the maximum allowable. Prison terms are also rising, but still, only one out of four people convicted of Clean Water Act criminal violations go to jail, and for trivial amounts of time. For corporations, criminal fines have tripled since 1983 but remain a tiny fraction of those allowed under the law. For example, in 1989, about two-thirds of all Clean Water Act corporate fines were less than half of the maximum allowed for a single day of violation.[157]

For perspective, it is useful to compare these sanctions with penalties for other types of crime under guidelines issued by the U.S. Sentencing Commission. For example, a pickpocket can be sentenced to four to ten months in jail, and a mugger can spend up to fifty-seven months in jail, pay fines of up to $100,000, or both. By comparison, when Ashland Oil spilled more than .5 million gallons of oil into the Ohio River, causing more than 1 million people to go without water for a week and causing the closing of schools, churches, and businesses in sixty communities, no individuals were indicted, and no jail time was served. Instead, the company pleaded no contest to a single count of violating the Clean Water Act and to one count of violating

TABLE 5.4

Average Clean Water Act Criminal Sanctions[158]				
INDIVIDUALS				
YEAR	AVERAGE FINE	AVERAGE PRISON SENTENCE	AVERAGE PRISON / % WHO GO TO JAIL AFTER SUSPENSION	AVERAGE PROBATION / % WHO GET PROBATION
1983–1984	$1,586	2.6 months	0 months / 0%	5.5 months / 60%
1985–1987	$1,240	1 month	1 day / 22%	6 months / 88%
1988	$4,880	27 days	27 days / 25%	5.3 months / 88%
1989	$ 793	1.5 months	2 days	4.6 months

CORPORATIONS		
YEAR	AVERAGE FINE	PERCENTAGE OF FINES LESS THAN:
1983–84	$10,750	$12,000 = 85%
1985–87	$18,862	$15,000 = 51%
1988	$29,760	$25,000 = 88%
1989	$34,375	$25,000 = 63%

the Refuse Act and paid a fine of $2.5 million, compared to $90 million in profits for the year. This was despite Ashland's long history of ignoring environmental statutes throughout the country; it had a prior conviction for violating twenty-two counts under the Clean Water Act.[159]

Unfortunately, the weaknesses in the Clean Water Act enforcement program are systemic problems rather than isolated breakdowns. For example, in two separate studies, EPA's inspector general found widespread problems with the enforcement of NPDES programs in EPA Regions III and IV.[160] The Region IV report found that enforcement programs in Georgia and Alabama were inadequate because

1. they lacked adequate enforcement management systems;
2. enforcement actions against chronic violators were not taken in a timely fashion or escalated when appropriate;

3. appropriate civil and criminal penalties were not used effectively; and
4. as a result, compliance with Clean Water Act deadlines was poor.[161]

Similarly, the Chesapeake Bay Audit (Region III) found that EPA did not require states to apply proper management procedures, that states were reluctant to take appropriate enforcement actions, and that EPA did not take enforcement actions when states failed to do so.[162] Similarly, an audit of EPA Region VI's handling of oil and gas discharges in coastal Louisiana found that "EPA enforcement under sections 402 and 404 of the Clean Water Act against oil and gas activities operating in coastal Louisiana is nonexistent." The report found that virtually all of these discharges were unpermitted.[163]

Finally, while section 505 of the Clean Water Act allows citizens to fill in the gaps when government enforcement actions are weak or absent, court decisions and government policies have weakened this "citizen suit" authority considerably. Most important, in 1987, the U.S. Supreme Court ruled that citizens cannot sue for "wholly past" violations of the law, requiring citizens to meet the higher burden of showing that violations are or may be continuing just to initiate a lawsuit.[164] This decision has caused a long series of problems for citizen enforcers. And many other barriers have hampered citizen enforcement efforts, some at the behest of the Department of Justice. These include federal attempts to wrest penalties away from cleanup efforts and into the federal treasury, and government efforts to preempt citizen suits with their own cases, which may result in inadequate penalties or cleanup.[165]

Conclusion

Taken together, the combination of poor effluent guidelines and pretreatment standards, incomplete water quality standards, inadequate permit limits, and poor enforcement of applicable permits yields a Clean Water Act program for point source dischargers that is a mere shadow of what Congress intended in 1972. These problems stand in the way of further progress toward achieving the fundamental goals of the Clean Water Act.

Chapter 6
Virtually Nonexistent Poison Runoff Controls

Poison runoff impairs more water bodies, surface and ground, urban and rural, than any other pollution source in the country. Poison runoff is the contaminated storm water and snowmelt that runs off of, or leaches through, land used and abused for human purposes without regard to ecological needs. Although poison runoff (nonpoint source water pollution) was widely acknowledged as a water quality problem even before 1972, in general we have failed to create and implement effective programs that protect and restore our nation's waters that are subject to this threat.

The framers of the 1972 Clean Water Act explicitly recognized the need for state water quality programs to address land-based sources of water pollution in their water quality assessments and in watershed management plans developed under section 208 of the act (called 208 plans). The dominance of the point source challenge, however, eclipsed public awareness of, and government attention to, more diffuse pollution sources.

By the mid-1980s, impatient with the lack of EPA and state progress in controlling poison runoff, Congress created the State Nonpoint Source Management Program (section 319). Unfortunately, the state 319 programs have been plagued by slow and inadequate funding, lack of adequate implementation mechanisms, and insufficient direction and oversight from EPA. The 1987 CWA Amendments also included requirements for the municipal and industrial stormwater permits previously discussed; these permitting programs are now helping to revive public interest in restoring blighted urban watersheds. In 1990, Congress passed a new program, aimed at reducing poison runoff in coastal watersheds, with a more ambitious pollution reduction mandate and more regulatory clout than the 319 program. The Coastal Zone Nonpoint Pollution Control Program may be a model for revisions to state runoff and watershed management programs that will help to reduce and prevent poison runoff.

To underscore the severity of the poison runoff problem, and to explain it to the uninitiated, we begin this section with a poison runoff primer. Next we evaluate the implementation and efficacy of Clean Water Act poison runoff programs that existed before 1987, as well as two major initiatives, one passed in 1987 (section 319) and the other in 1990 (as part of the Coastal Zone Management Act).

A Primer on Poison Runoff

Poison runoff is not a new phenomenon; in fact, it has been with us since the first settlers clear-cut the New England forests and since the first farmers began plowing the fertile lands of the eastern coastal plain. Reflecting an absence of understanding of history, the official rhetoric has often apologized for the severe lack of money, staff resources, and regulatory clout needed to reduce poison runoff by claiming that this is a new or obscure pollution source. For example, a 1989 EPA report to Congress on section 319 of the Clean Water Act states,

> Nonpoint source impacts have not been fully assessed. The Nation has focused largely on impacts caused by traditional point sources (POTWs and industrial dischargers) in the past because point source discharges were causing major, visible problems in our surface waters. Thus, very little attention has been given to assessing the impacts of NPSs.... Since water quality impacts still exist in many areas, it is now very clear that NPSs have had and continue to have widespread impacts upon surface waters.[1]

Contrary to this assertion, framers of the original Clean Water Act well knew about land-based, diffuse pollution sources and the severity of damage they caused. The 1972 Senate report said:

> One of the most significant aspects of this year's hearings on the pending legislation was the information presented on the degree to which nonpoint sources contribute to water pollution. Agricultural runoff, animal wastes, soil erosion, fertilizers, pesticides and other farm chemicals that are a part of runoff, construction runoff and siltation from mines and acid mine drainage are major contributors to the Nation's water pollution problem. Little has been done to control this major source of pollution.... It has become clearly established that the waters of the Nation cannot be restored and their quality maintained unless the very complex and difficult problem of nonpoint sources is addressed.... The Committee recognizes, at the outset, that many nonpoint sources of pol-

lution are beyond present technology of control. However, there are many programs that can be applied to each of the categories of nonpoint sources and the Committee expects that these controls will be applied as soon as possible.[2]

Unfortunately, it would be more than two decades before any land use category-specific water quality controls would be required as part of a federal program—the Coastal Zone Nonpoint Pollution Control Program, which we will discuss later. In the intervening years, poison runoff continued unabated.

Widespread Poison Runoff Damages

Two water quality assessment programs required by the CWA cover poison runoff: the biennial 305(b) reports and the one-time 319(a) reports. The 305(b) reports are supposed to cover all water bodies and all relevant pollution sources in each state; the 319(a) reports are supposed to be statewide assessments of runoff problems, conducted wherever possible on a watershed-by-watershed basis. There is some overlap between these two reports.

As we discussed earlier, the 305(b) water quality assessments are difficult to compile for a time-series analysis of trends, since the scope and methodologies for reporting have changed so frequently. And these reports likely underestimate the magnitude of poison runoff even more than other sources of pollution, because, as we will discuss, physical and biological (as opposed to chemical) impairments dominate poison runoff even more than they do point source pollution. The most complete, and thus the most revealing, 305(b) reports on runoff problems were from EPA's most recent reporting cycle (1988–89).

In 1991, EPA published a compendium of the states' 319(a) assessments, entitled *Managing Nonpoint Source Pollution,* as required by section 319(m). This report also contains a comprehensive set of statistics on the role of land-based sources in damaging aquatic resources nationwide. Following, we summarize the damage assessment from this report as well as from the 1988–89 305(b) compilation (the *National Water Quality Inventory*).

Rivers. Agricultural runoff impairs or threatens more than 100,000 assessed river miles nationwide. Logging impairs more than 15,000 more assessed river miles, and construction runoff impairs almost 10,000 assessed river miles. The 319(a) reports list about 40,000 river miles as threatened by runoff pollution sources.

Lakes. Agricultural runoff impairs almost 2 million acres of U.S. lakes. Storm sewers impair almost another million acres.

Great Lakes. Poison runoff contributes to the impairment of designated uses in portions of Lake Michigan in Indiana (protection of wildlife) and parts of Lakes Erie and Ontario in New York (protection of fisheries). (No other Great Lake state provided quantitative assessments of runoff impacts to the Great Lakes.)

Wetlands. About 52,000 acres of wetlands in California, Iowa, and Delaware are not supporting one or more designated uses or are threatened by poison runoff sources. (No other states gave quantitative information on wetlands damage from runoff sources.)

Coastal Waters. Due to runoff, 1.2 million acres of coastal waters are not fully supporting one or more designated uses.

Estuaries. Runoff sources impair or threaten about 5,000 square miles of estuarine waters.

Groundwater. Runoff sources threaten public drinking water supplies in the four states that specified impacts to designated uses.[3] Nitrates in groundwater exceed current health standards in virtually all states and occur in 5 to 20 percent of sampled wells in the western Corn Belt and Mid-Atlantic states, largely due to fertilizer applications on farms.[4]

The runoff management and water-body assessment programs are not the only sources of national statistics on runoff damage. In June 1989, under section 304(l), discussed previously, EPA released a list of more than 17,000 toxic hotspots—seriously degraded water bodies. Factories or sewage treatment plants impaired only 602, or less than 4 percent, "wholly or substantially." The rest were polluted, wholly or substantially, by poison runoff from farms and other sources.[5]

User-Friendly Watershed Health Reports

These general statistics, of course, only tell us about part of the problem. As EPA points out, the states vary widely in their format for reporting on runoff problems; not all states reported on all important runoff sources, and not all reported on watersheds as well as on water bodies.

When nurses walk onto a hospital ward to begin their daily rounds, they look first for each patient's medical chart, which provides, at a quick glance, a snapshot of that person's current state of health. Water activists and managers require a similar watershed health status report that provides a quick glance at each watershed's pollution problems, sources, and remedies recommended and applied. Ideally, 305(b) reports—as well as 319(a) reports—would provide such watershed health reports; unfortunately, many do not.

Oregon's 319 assessment, however, is exceptional; it gives detailed and readable summaries of the health of all eighteen river basins in the state, on a whole-watershed basis. Oregon's Department of Environmental Quality (DEQ) created these summaries and maps by using an extensive network of public and professional resource experts and a Geographic Information System (GIS). The DEQ assessed 27,722 river miles—roughly one-quarter of the state's total 112,600 miles—for runoff effects. Figure 6.1 depicts the seven most severely impaired river basins in Oregon. All have "severe impairment" in 30 percent or more of assessed river miles.

In addition, Oregon identified the potentially responsible land uses, by percentage of total watershed area, for each watershed (see Figure 6.2, regarding the John Day River Basin). The most commonly cited causes of beneficial use

FIGURE 6.1

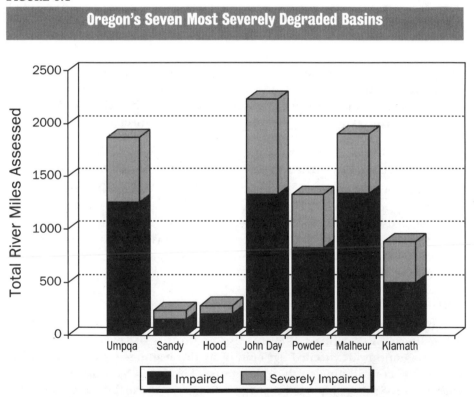

Source: Oregon DEQ, *Oregon Statewide Assessment of Nonpoint Sources of Water Pollution* (1988), 2 (table 1.1).

FIGURE 6.2

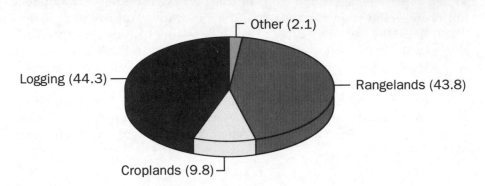

John Day River Basin

Other (2.1)

Logging (44.3)

Rangelands (43.8)

Croplands (9.8)

Source: Oregon DEQ, *Oregon Statewide Assessment of Nonpoint Sources of Water Pollution* (1988), 2 (table 1.1).

degradation were vegetation removal along streambanks, removal of thermal cover over streams, surface erosion, and changes in flow pattern and timing. The land uses most commonly cited in connection with these problems were grazing, recreation, irrigated and nonirrigated agriculture, and forestry.

The Nature of Poison Runoff by Land Use Category

Virtually every human activity on the land has the potential to impair water quality and aquatic habitat. It is beyond the scope of this report to describe every land use category in detail. We will, however, highlight the categories that do the most damage nationwide: agriculture (including cropping, grazing, and livestock confinement), mining, urban development, and logging.

Agriculture

Agriculture is the leading source of water pollution in the United States, according to EPA.[6] In its 1989 release of the list of seventeen thousand hotspots nationwide, it cited agriculture as the leading source of pollutants causing or contributing to those hotspots.[7] The 1990 *National Water Quality Inventory* (1988—89) reported that agriculture was by far the largest source of river impairment, contributing to more than 60 percent of impaired river miles. For perspective, the next biggest reported source—municipal sewage plant discharges—contributed to 16.4 percent of impaired river miles.[8]

Agriculture is also a leading cause of species endangerment and extinction. About 37 percent of the 436 species listed in the Endangered Species Information System database are imperiled at least in part by irrigation and the use of pesticides. An unpublished EPA staff report of November 1989, based on data from the Department of the Interior, identified 125 endangered or threatened species that are aquatic or water dependent and that agricultural practices such as pesticide usage affect.[9]

Agricultural activities were also fingered as a major cause of fish kills. Three of the top six pollutant categories cited as causing fish kills—low dissolved oxygen, pesticides, and fertilizers—are wholly or substantially from agricultural uses. (The other three are petroleum, pH/acidity, and organic chemicals.) However, agriculture accounted for only 5 percent of the total number of fish killed from 1977 through 1985, because the size of each fish kill was relatively small. In EPA's 1986–87 summary of state reports on fish kills, animal feedlot and waste operations were blamed for more than 1 million fish killed (most likely due to oxygen starvation from manure pollution).[10] In a separate 1984 survey of fish kill data, pesticides were cited as the leading documented cause of fish kills in the United States over the previous two decades.[11]

Because agriculture is by far the biggest source of water-body impairment nationwide, and because it is such a diverse industry, we must subcategorize the industry in order to explain regional differences in the types of impairments observed.

Cropping

Soil erosion, pesticide pollution, nitrates leached into groundwater, nitrogen and phosphorus runoff into estuaries, wetlands conversion, streambank wastage, and manure runoff are all major problems associated with crop production. Irrigated crop production can be associated with all of these water quality problems, as can the discharge of toxic mineral salts into estuaries and marshlands. Following are some national and regional data on water pollution from crop production.

Soil Erosion. Data has been collected on the national level only since 1977. The *National Resources Inventory*, taken roughly every five years by the Soil Conservation Service, includes reports for 1977, 1982, and 1987. The 1982 and 1987 reports, which are more reliable than the 1977 reports, show a decline from roughly 1.8 billion tons of sheet and rill erosion in 1982 to about 1.6 billion tons eroded in 1987—11 percent.

Among the trends in crop production accounting for the decline is the onset of the 1985 Farm Bill conservation program, including the conservation reserve and conservation compliance programs. While a precise connection between soil erosion and water quality cannot be made, in general, the more soil is conserved, the more our waters are protected from sediment pollution.

Pesticides. These chemicals pollute both surface water and groundwater. We have already discussed fish kills from pesticides. Overall, pesticide use statistics are startling and indicate the magnitude of the potential pesticide problem for water quality. According to EPA estimates, approximately six hundred active ingredients are marketed in forty-five thousand to fifty thousand formulations. About 430 million pounds of pesticides were applied agriculturally in 1987, with a market value of about $4.0 billion.[12] According to EPA's compilation of the states' 1988–89 305(b) reports, pesticides impaired 11.2 percent of all assessed river miles and 14.5 percent of Great Lake shore miles. (This compilation did not include an assessment or report of pesticide impairment of lakes.)[13]

A recent GS study of ten current-use herbicides in surface waters of the Midwest found that high concentrations of herbicides were flushed from cropland and were transported through surface waters as pulses in response to late spring and early summer rainstorms.[14] Several of the herbicides exceeded EPA water quality criterion for drinking water–human health protection in a significant percentage of the samples. For example, 52 percent of the sites exceeded the primary drinking water standard ("maximum contaminant level," or MCL) for atrazine (3 micrograms per liter, or ug/L); 32 percent exceeded the MCL for alachlor (2 ug/L); and 7 percent for simazine (1 ug/L). The median concentrations of the four major herbicides—atrazine, alachlor, cyanazine, and metolachlor—jumped by a factor of ten in the late spring–early summer samples and then dropped back to near-preplanting levels by harvest time. The study sampled 149 sites in 122 river basins of Ohio, Indiana, Illinois, Wisconsin, Minnesota, Iowa, Kentucky, South Dakota, Kansas, Nebraska, and Missouri.[15] The fact that more than half of these midwestern surface water sites exceeded the atrazine drinking water standard, and one-third exceeded the alachlor drinking water standard, is a concern for all communities that rely upon these waters for drinking water, and particularly for those small rural towns that do not use advanced drinking water treatment such as carbon filtration.

Phosphorus and Nitrogen. Agricultural operations, including the spreading of manure and application of artificial fertilizer, disseminate these important water pollutants. Phosphorus in high levels is acutely toxic to fish; in much lower levels, it overenriches water bodies, causing them to fill up with algae through eutrophication. Nitrogen, especially in the form of nitrate, is a human health and livestock health concern (EPA's drinking water standard (MCL) for nitrate is 10 milligrams per liter, or mg/L) because it causes "blue baby syndrome" (methemoglobinemia). Like phosphorus, nitrogen contributes to eutrophication of lakes and estuaries even in much smaller concentrations than those of human health concern. In the form of ammonia, nitrogen is also acutely toxic to fish.

The use of nitrogen fertilizers in the United States increased by more than a factor of four in the two decades between 1960 and 1981, to a 1981 total of 11.9 million tons per year. Per-acre use of fertilizers doubled between 1964 and 1984. However, the use of nitrogenous fertilizers (n-fertilizers) may have reached its apex in the 1980s and has apparently begun to decline slightly; the total tons of n-fertilizer use declined 12 percent, to 10.5 million tons of annual application, between 1981 and 1988.[16]

Long-term trend data for nitrate pollution is scarce. One ten-year study in Nebraska from the early 1960s to the early 1970s showed a 25 percent increase, on statewide average, of groundwater nitrate-nitrogen concentrations. During that same period, nitrogen fertilizer use in Nebraska increased by a factor of four. A longer-time-series study on nitrogen pollution, based on almost forty-six hundred samples from wells all over Iowa, showed that nitrate levels in groundwater from shallow wells less than 100 feet deep increased slowly but steadily from 1952 to 1979 where total fertilizer use and per-acre applications were increasing rapidly.[17]

Waters in karst (limestone) topographies are especially vulnerable to nitrate contamination. In Iowa's Big Spring Basin, part of the karst region that straddles portions of Iowa, Wisconsin, and Kansas, groundwater nitrate concentrations tripled, from 5 mg/L to 15 mg/L, from 1958 to 1982. These data suggest a yearly rate of increase of 0.4 mg/L of the average nitrate-nitrogen concentration. A farm survey in the basin in the mid-1980s showed that area farmers were not preparing nitrogen budgets to determine appropriate fertilizer application rates. Where such budgets were prepared, they were incomplete. Alfalfa and manure contributions to soil nitrogen were being neglected, and thus, in 1984, artificial fertilizers were being applied in excess of need at a rate of about 90 kg/ha.[18]

The foregoing examples of nitrate contamination trend data are perhaps from regions with vulnerable climate and geology, and the problems of nitrate contamination inevitably vary in severity from region to region. Nonetheless, "snapshot" statistics from single-year national studies show that nitrate contamination of groundwater is indeed a problem of national scope. A 1986 GS sampling of 316 principal aquifers in forty-six states turned up 288 (91 percent) with median nitrate-n levels below 3 mg/L; twenty-seven (8.5 percent) with median nitrate-n levels between 3 and 10 mg/L; and one (0.5%) with a median nitrate-n level above 10 mg/L. The same GS study found that forty-one aquifers in twenty states (13 percent) had nitrate hotspots where more than 10 percent of the samples exceeded the EPA–human health 10 mg/L standard.[19]

Agricultural nitrates are also a problem for some surface waters. Although GS did not detect any significant trends in nitrate increases in a majority of the 383 river and stream stations it monitored nationwide between 1974 and 1981, nitrogen concentrations increased at 116 stations (by 13 percent) during that time. Atmospheric deposition of nitrogen apparently plays a large role in the frequent occurrence of nitrate increases in rivers of the Midwest and Mid-Atlantic regions.[20]

Irrigation Pollution. Irrigation agriculture, which accounts for 90 percent of the water consumed in the West, results in poisoned return flows that cause serious damage to waters and wetlands, endangering aquatic wildlife with toxics, including selenium, boron, molybdenum, and chromium. Selenium has been identified as the cause of an observed high rate (64 percent) of deformed and dead bird embryos at Kesterson National Wildlife Refuge in California.[21]

Although irrigation flows begin with the diffuse flow of irrigation water off of farm fields, they end as point source discharges, conveyed through pipes or ditches. The Clean Water Act gives irrigation return flows express exemption from NPDES permitting, without any database showing that the flows are benign. In fact, the FWS's preliminary data indicate that almost half (48 percent) of its refuges that have toxic contaminant problems receive agricultural drainage.[22]

Grazing

Accurate national statistics on the total water quality damage wrought by grazing on both public and private rangelands are not currently available. However, the surveys that have been conducted on public rangelands do show massive damage to riparian areas from overgrazing. Statewide surveys by the Bureau of Land Management (BLM) in Colorado and Idaho, and more limited BLM surveys in Nevada and Utah, showed that more than 80 percent of assessed streams or riparian areas were in poor or fair condition. Surveys by the U.S. Forest Service (USFS) produced similarly troublesome results; in Arizona, 80 to 90 percent of the stream riparian areas in Tonto National Forest were in unsatisfactory condition.[23]

Rangeland expert Lynn Jacobs gives additional data on grazing damages to streams in the West, citing wildlife ecologist Charles Kay: "A recent study in

Wyoming found that of 262 miles of streams, only 2% function now as they did in 1850. Eighty-three percent of the streams were lost or destroyed by overgrazing and accelerated erosion. The remaining 15% were in fair to good condition."[24] According to another range technician, riparian damage from cattle is so widespread in the West that most people, including most range managers, have never seen a healthy stream channel.[25]

This riparian damage to vast areas of the West is for the sake of a tiny proportion of the nation's livestock. Although 90 percent of our western BLM lands are used for ranching, they produce only about 1.1 percent of U.S. cattle and sheep.[26] A GAO report on the health of riparian areas on U.S. public rangelands points out that the preferred management practice, cattle exclusion from streamside zones combined with revegetation, can reduce many of these impacts. Unfortunately, many areas do not require this practice, and BLM staff are thwarted by their own top management in carrying out riparian restoration projects.[27]

Livestock Confinement (Feedlots)

EPA's Office of Policy, Planning and Evaluation (OPPE) estimates that, based on the U.S. Census of Agriculture, at least 1.1 million farmers have livestock. Of those, only five thousand to ten thousand operations nationwide may be above the current thousand-unit cutoff for NPDES permit issuance.[28] For the rest, no particular federal manure management requirements apply under the Clean Water Act. As an example of the severity of the manure pollution problem, a Chesapeake Executive Council report noted that "[c]ontrol of 85 percent of Pennsylvania's animal waste alone would accomplish a 40 percent nutrient reduction for the state."[29] The Chesapeake Bay Foundation concluded that Pennsylvania should toughen its manure management program, including the "targeting of enforcement efforts at those operations responsible for disproportionately high nutrient loads as well as committing more resources to the program in general."[30]

Mining and Resource Extraction

Of the 171,008 impaired river miles assessed by the states for the 1988–90 305(b) reports, 14 percent, or almost 25,000 miles, were polluted by mining runoff (designated "Resource Extraction" by EPA.)[31] As reported by GAO, a 1976 study by an EPA contractor found that "80 percent of the nonpoint source pollution from inactive and abandoned ore and mineral mining areas was occurring in five states—California, Colorado, Idaho, Missouri, and Montana.... The principal pollutants from these mines and mine waste piles were acid mine drainage, heavy metals, and sedimentation."[32] As with all poison runoff sources, assessed sites are only a small portion of the total; Colorado, for example, had studied the environmental impact from only about one-sixth (eight thousand of an estimated fifty thousand) of the state's noncoal abandoned or inactive mines.[33]

Urban Development

Runoff from construction impaired more than 9,800 river miles, or 5.7 percent of total impaired miles, in the 1988–90 reporting cycle, and storm sewers from urban sites impaired more than 18,000 river miles, or 10.6 percent of total impaired miles, in the same cycle. (Urban watershed degradation and restoration are discussed at length following.)[34]

Logging

States reported that a total of 9 percent of impaired river miles, or 15,459 miles, were polluted by silvicultural activities in the 1988–90 reporting cycle. These figures are probably gross underestimates, however, as some key logging states, such as Maine, do not monitor logging-related parameters such as siltation levels.[35] As EPA points out in its final 319 report, "The absence of information from 12 states significantly distorts the figures; Alaska and Oregon, in particular, have considerable forestry activity and their inclusion would have affected the total."[36]

Fisheries biologists in the Northwest have discovered that logging tends to destroy fish habitat more profoundly than previously believed. Siltation from logging operations has long been known to clog the gravel beds that are the spawning grounds for threatened salmon species. Only since the early 1980s, however, have biologists discovered that salmon survival requires more than silt-free gravel beds for spawning. It also requires extensive drainageway protection, since the young of the year are reared in tiny, capillary-level, first-order tributaries, abandoned meander ponds, and seep-fed creeks. According to naturalist Robert Steelquist:

> There they grow rapidly on aquatic insects and other organisms. This burst of growth gives these cohos a distinct advantage for survival at sea when they eventually leave the freshwater system.... These pond and tributary habitats, however, had never been recognized for their contribution to coho productivity. Though measures were in place to protect main-stem habitats from destruction, the beaver ponds and small channels were particularly vulnerable to logging, road building, and culverts, often filling with slash and debris.[37]

The road cuts, skid trails, and clear-cuts that timber companies bring to forestlands do extensive damage to streams, rivers, and lakes around the country. In a study done in the late 1970s, for example, the Maine Forest Service found the following: 52 percent of harvesting sites near state-designated protection zones had erosion or sedimentation problems; a substantial number of sites violated logging road runoff, and stream-crossing requirements; and siltation was a problem in water bodies 75 to 250 feet from logging operations.[38] Despite this evidence of widespread harm to water quality, water-sensitive practices for logging sites are still voluntary for the vast majority of the Maine Woods.

Runoff Mandates Abandoned or Poorly Implemented

The Clean Water Act has addressed runoff pollution explicitly since the act's inception in 1972. As NRDC points out in the book *Poison Runoff,* runoff control mandates in the pre-1987 Act could have been used more effectively. In fact, since 1972, the CWA has required that EPA and the states devise comprehensive programs to control water pollution from both point and nonpoint sources. At least five pre-1987 sections of the act—102(a), 201(c), 208, 303, and 305(b)—relate to, or explicitly describe, poison runoff assessment, control, and reduction.[39] Following, we describe briefly these requirements of the original Clean Water Act and the degree to which they were implemented around the country.

Sections 102(a), 201, and 208 provided broad authority to EPA to set up holistic pollution prevention programs to protect water quality (long before *pollution prevention* became a popular term). Section 201(c), addressing areawide waste treatment management, was designed to ensure that state and local managers of the construction grants program would not have "point source tunnel vision." Congress wanted comprehensive water pollution benefits through the control or treatment of *all* pollution sources, not just point sources of raw sewage and industrial waste.

Section 208 can be seen as further explication of the comprehensive program goal set forth in section 201. Section 208 is perhaps the best known of the act's pre-1987 poison runoff requirements, partly because so many citizens participated in the creation of 208 plans. Section 208(b)(2)(F) requires areawide waste treatment management plans to include:

a process to (i) identify, if appropriate, agriculturally and silviculturally related nonpoint sources of pollution, including return flows from irrigated agriculture, and their cumulative effects, runoff from manure disposal areas, and from land used for livestock and crop production, and (ii) set forth procedures and methods (including land use requirements) to control to the extent feasible such sources.

A series of congressional hearings in 1979 highlighted the following problems as having hindered the success of the 208 program:

- too little time in which to create the plans;
- discontinuity and lack of federal funding;
- inadequate water quality data; and
- poor management by EPA.[40]

These same hearings emphasized several obstacles preventing managers from implementing practices to stem the flow of runoff:

- inadequate data on the effectiveness of control measures;
- institutional conflicts;
- lack of public education on the benefits of nonpoint source control;
- inadequate funding; and
- debates over regulatory versus voluntary approaches to induce cooperation.[41]

A total of 176 section 208 plans were created, plus another 49 state-level areawide plans. These 225 comprehensive water quality plans represented a definite step forward in the national knowledge base on diffuse, land-based pollution sources and on watershed management in general. Another strength of the 208 process was that it involved high levels of participation from the public, particularly citizen leaders from the League of Women Voters and from local resource conservation districts.[42]

Sadly, during the 1980s, most 208 plans were shelved, and their excellent concepts fell by the wayside. In addition to the overall lack of implementation mandates and other administrative problems already listed, reasons included lack of funding; EPA timidity in issuing stringent guidelines and in linking 208 implementation with mandates to achieve water quality standards;[43] and turf battles that flared up when watershed boundaries cut across political boundaries.

Earlier, we described sluggish progress in employing a number of other tools, basic to the CWA, that have potential power to stem the flow of runoff. These include water quality standards, the 303(d) TMDL approach, 305(b) assessments, and other basic CWA tools. Water quality standards and their implementing mechanisms, including effective state antidegradation programs, are especially important to the success of runoff reduction programs and are crucial to runoff programs for two basic reasons:

1. All programs to control poison runoff must be designed to achieve compliance with water quality standards;[44] and
2. At least before 1987, water quality standards formed the principal legal authority for controlling pollution generated by various land use activities.

Water quality standards must work hand in hand with well-targeted monitoring and assessment programs in order to be effective for any pollution control program, including runoff control. Section 305(b)(1)(E) of the Clean Water Act requires that the biennial state water quality assessments include "a description of the nature and extent of nonpoint sources of pollutants, and recommendations as to the programs which must be undertaken to control each category of such sources, including an estimate of the costs of implementing such programs."[45]

Nonetheless, an EPA report on the main elements of state water pollution source monitoring programs suggests that, at least as of 1987, they were characterized by point-source "tunnel vision": (1) self-monitoring of effluent by industrial and municipal dischargers, (2) compliance sampling inspections to cross-check discharger self-monitoring, and (3) effluent characterization studies for industrial dischargers. One of the five major challenges set forth for EPA in this study is to "Identify and Characterize Toxic, Conventional, and Anthropogenic Pollutants from *Nonpoint Sources*" (emphasis added). This report also recommended an in-depth study of the feasibility of initiating a citizen's watch program. Both of these recommendations, if EPA followed them, would significantly help states to characterize threats and impairments due to land-based sources of pollution.[46] The good news is that, apparently, many states are now beginning to shift their monitoring efforts into land-based sources of water pollution, at least according to one 1992 report described in the next section.

The early 1980s represented perhaps one of the lowest periods in the history of poison runoff policy. Funding for the 208 program was gutted in 1981.[47] Then, in 1983, EPA contended that it had no direct role in controlling poison runoff. In addition, the Reagan administration actively opposed the establishment of a new, comprehensive runoff control policy.[48] Obviously dissatisfied with the lack of progress made by states in stemming the flow of runoff, Congress created new requirements in the 1987 Clean Water Act. For the first time, poison runoff was addressed head-on in a new section of the act.

Increased Emphasis on Runoff Programs in 1987

In 1987, Congress created section 319 of the Clean Water Act to get states to identify waters damaged or threatened by runoff sources and to develop comprehensive programs to heal those waters by reducing and eliminating pollution from those land-based sources. This program was not completely new; it corralled provisions for runoff controls dispersed throughout the act and EPA guidance.

This section of the act strengthened the substantive standard for runoff control program effectiveness by requiring, in 319(a)(1)(C), nonpoint source reduction "to the maximum extent practicable." By contrast, the earlier 208 programs had held to a much weaker standard of runoff reduction "to the extent feasible." As NRDC observed in *Poison Runoff,* the new standard "will demand a higher level of control and a more stringent standard of proof before degradation or downgrading can be permitted."[49] Unfortunately, the promise of section 319 has not been fulfilled. True, there have been some notable success stories, described following. On a national basis, however, significant progress has not been shown under the new program.

Watershed Restoration Success Stories

Although Congress intended for the states to structure their poison runoff control programs as much as possible on a watershed basis (319[b][4]), many states did not do so, choosing instead to write management plans based upon generic management practices intended to apply to all lands within each major land use category in the state. Some states, like Wisconsin, are exceptional in that they include runoff control as part of a comprehensive, watershed-based restoration and protection program that targets specific watersheds throughout the state. Other states are notable for individual watershed programs that stand as examples for others to follow. Descriptions of three such programs follow.

Owl Run Watershed, Fauquier County, Virginia. A major nutrient and animal waste problem here contributes to dissolved oxygen and other water quality problems in the Chesapeake Bay. In response, the Virginia Division of Soil and Water Conservation, in cooperation with conservationists at the John Marshall Soil and Water Conservation District, is helping farmers to reduce manure pollution through a variety of techniques. Management practices in the 2,800-acre watershed include soil testing and the creation of nutrient budgets; no-till cropping and filter strips; and the construction of manure storage tanks. Cost-sharing can cover up to 100 percent of the farmer's installation expenses. The project is intended to show that these kinds of practices are effective on a whole-watershed basis in reducing pollution. The water quality goals around which the project is designed include both instream and downstream (Chesapeake Bay) restoration.[50]

Big Darby Creek Watershed, Ohio. This multiparty project in central Ohio, coordinated by the Nature Conservancy (TNC), proves that farmers and environmentalists can be friends. A major goal of the project is to enroll 75 to 100 percent of the watershed's farmers in a conservation tillage program; roughly 15 percent of the watershed's farmers now use conservation tillage. Big Darby is a unique ecosystem with many endangered or threatened species of mussels and fish. Enhancing cooperation and rapport are now-famous canoe trips in which a farmer and an environmentalist survey the riparian zones together as they glide down the river. Many farmers in the 370,000-acre watershed are also installing forested buffer strips with the help of state foresters. The project's 1992 budget totaled more than $750,000, with monies obtained from the SCS, TNC, EPA, and other agencies and groups. Big Darby is not a purely agricultural watershed. A remaining wild card for the fate of its headwaters is whether surburban developers, seeking to supply wealthy residents of Columbus with low-density country housing, will be convinced to adopt water-sensitive practices of their own.[51]

Big Spring Basin, Iowa. The Iowa Department of Natural Resources, Geological Survey Bureau, has helped to make the Big Spring Basin famous for nitrogen input reductions that have saved farmers money while reducing water pollution. Through a state cost-sharing program and extensive technical outreach to the roughly two hundred basin farmers, a reduction of more than 1.2 million pounds of applied nitrogen was achieved between 1981 and 1989. This input reduction achieved a savings of about $200,000 per year, or an average of $1,000 per year per farm. With crop rotations that have farmers planting corn following alfalfa, maximum yields are often obtained without any nitrogen added to the soil.[52]

These watershed success stories are cause for hope that whole-watershed restoration works, that cooperation between different stakeholders is possible, and that farmers are willing to adopt water-sensitive practices once they are convinced of three things: (1) that an ailing or vulnerable ecosystem needs such changes; (2) that such changes will not bankrupt their farms (and may even save them money); and (3) that all other farmers and landowners in the watershed share equally the risk and burden of adopting new practices.

The Owl Run, Darby Creek, and Big Spring Basin examples are voluntary programs with the premise that, given ample time, money, and technical outreach, all farmers will volunteer to do the right thing. Unfortunately, states may not support these types of programs in all impaired or threatened watersheds, since ample grant monies to replicate their favorable cost-share ratios statewide simply do not exist. The needs for urgent action in the case of impaired watersheds and for accountability demand more than voluntary programs.

Recent Federal Oversight Lacking in Vision and Leadership

The EPA's OPPE reviewed the 319 program in the summer of 1992 at the request of the Office of Water. It looked at ten sample state programs and at management policies at the EPA headquarters and regions. The report reached twelve findings about what's right—and what's wrong—with the 319 program:

1. Because of the diverse nature of nonpoint source (NPS) pollution, there is no single definition of an NPS program.
2. Authority for implementing state management programs is generally decentralized.
3. The extent to which states are institutionalizing their NPS programs varies widely.
4. The majority of the ten states do not have NPS programs oriented toward improving water quality on a watershed-specific basis.
5. State management programs generally cannot be used to gauge the states' progress in implementing NPS controls.
6. Flexible guidance has enabled states to use 319 resources to address many NPS priorities.
7. States concentrate their use of 319 resources to focus on different priority activities.
8. The majority of states are making some effort to monitor the effectiveness of best management practice (BMP) implementation, though water quality impacts due to 319 implementation are not yet known.
9. Section 319 has facilitated increased communication and coordination among agencies and organizations to develop and implement the state management programs.
10. Although most EPA regional offices use several staff to address NPS pollution, few staff are dedicated specifically to assisting states in implementing management programs or 319 grants.
11. Office implementation of the 319 grant program varies considerably across EPA regions.
12. Although EPA provided states the opportunity to develop diverse NPS programs, it has not yet defined a vision or role for a national NPS program.[53]

The Office of Policy, Planning and Evaluation then made the following two recommendations: (1) the Office of Water should emphasize more clearly that a watershed protection approach should be the basis of state NPS programs and (2) the Office of Water and regional offices should clearly define EPA's goals, strategy, and role for the national NPS program.[54]

One of OPPE's most important findings was that "[t]he majority of the ten States do not have NPS programs oriented toward improving water quality on a watershed-specific basis." Furthermore, "the majority of [state management programs] do not identify strategic plans or milestones for achieving water quality goals for specific waters identified in their Assessment Reports."[55] Thus, EPA simply has not enforced the requirement of section 319(b)(4) that states shall, to the maximum extent practicable, develop and implement management programs on a watershed-by-watershed basis. Although some state-to-state variation is expected and even desirable in the 319 programs, the report clearly suggests the need for more program focus at both the federal and the state levels.

Adequate Funds Lacking

The GAO also reviewed section 319 program implementation in 1990. It found that "officials in five of the states we visited identified the lack of resources as a key barrier to controlling nonpoint source pollution. Although some states have or will allocate millions of dollars to deal with the problem, they maintain that it would require billions to correct."[56]

The total 319 appropriation for the past four fiscal years—roughly $200 million—represents a drop in the bucket, compared both to present program needs and to the tens of billions of dollars the nation has invested in sewage treatment (significant, given that poison runoff pollution dwarfs the sewage treatment challenge of the early 1970s).

Findings on Runoff Control Programs, and Changes Needed

As discussed here and in previous chapters, there are major gaps in the development and use of the Clean Water Act's basic tools. The lack of implementation is perhaps most acute within the context of fledgling state poison runoff control programs, partly because the tool of NPDES discharge permits is usually not available to give these programs the backbone and bite of an automatic enforceable mechanism. Thus, the relative weakness and underdevelopment of the tools that *are* available to runoff managers—water quality standards, water quality assessments targeted to land-based sources, TMDLs, and whole-watershed plans—have hindered progress in stemming the flow of poison runoff.

New Water Quality Criteria Needed for Runoff

Although EPA took a quantum leap forward with the publication of its document *Biological Criteria: National Program Guidance for Surface Waters* (April 1990), few states have acted to use biocriteria in important ways in assessment and/or permitting. No water quality standards at the state or federal level have been established to protect physical or hydrological features of aquatic habitat, such as the first-order streams destroyed by logging, as discussed earlier. To protect whole aquatic ecosystems from the abuses of shopping mall and subdivision development, logging, mining, and other land operations, EPA needs to publish, and the states need to implement, the following water quality criteria:[57]

● biocriteria, such as EPA's recommended use of the Index of Biotic Integrity, first developed by Dr. James Karr and colleagues;

● habitat protection criteria—for example, for pool and riffle complexes;

● drainage density metrics, including minimal preservation and restoration of first-order streams;

● complete hydrologic specifications, including year-round flow minima and minimum stream flow percentages of groundwater;

● seasonal and annual sediment loadings;

● nutrients (for eutrophication, not acute toxicity); and

● current-use pesticides.

In its review of EPA's management of the overall poison runoff program, the GAO listed the lack of appropriate standards as a key barrier to progress:

> "Criteria documents" and other technical information are not available to states to enable them to set water quality standards for nonpoint source pollution.... State and federal officials told us that existing state water quality standards need to be supplemented because they were developed primarily to address point source problems and consequently have limited applicability in controlling nonpoint source pollution.[58]

The States' Poison Runoff Management Programs

In summary, the 319 program, as implemented over the past five years, has failed to heal waters and watersheds damaged by land uses and abuses for many reasons:

1. lack of watershed basis for the programs;

2. lack of adequate funding, especially for program staff at all levels;

3. inadequate enforcement of the mandate for states to require water-sensitive practices to be adopted wherever monitoring indicates a problem, or where pristine conditions indicate the need for protection;

4. major monitoring gaps;

5. inconsistent goals of other powerful federal programs that thwart poison runoff control efforts;

6. continued state reliance on ineffective voluntary compliance for landowners to adopt water-sensitive practices;

7. reluctance to create relevant water quality standards to make the program meaningful; and

8. diffuse responsibility for the program, which is often administered and overseen by agencies that lack a primary water quality focus.

With these major obstacles, our national poison runoff policy is based upon a voluntary, piecemeal approach riddled with inconsistencies, ineffectiveness, and massive gaps in funding, monitoring, and staffing. As a result, we now have fifty individual runoff assessment and management programs that vary widely in terms of comprehensiveness, stringency, degree of public participation, accountability, funding commitments, and instream effectiveness. And most programs fall on the voluntary side of the spectrum. Major changes are needed to strengthen 319 into a publicly accountable and ecologically and economically effective program.

The New Coastal Runoff Program—Promising but Limited in Scope

As part of the 1990 Coastal Zone Act Reauthorization Amendments (CZARA), Congress welded two existing programs—the states' Coastal Zone and Clean Water Act section 319 programs—into a single, powerful approach to preventing and reducing runoff pollution in coastal watersheds (including the Great Lakes). The centerpiece of CZARA is implementation of enforceable management measures to reduce polluted runoff by specific land uses. Management measures are defined in section 6217(g)(5) of CZARA as

economically achievable measures for the control of the addition of pollutants from existing and new categories and classes of nonpoint sources of pollution, which reflect the greatest degree of pollutant reduction achievable through the application of the best available nonpoint pollution control practices, technologies, processes, citing criteria, operating methods, or other alternatives.[59]

The phrase "greatest degree of pollutant reduction achievable" is more stringent than the "maximum extent practicable" standard for BMPs under section 319. Other important provisions of CZARA include:

- the extension of coastal zone boundaries farther inland, to control the land and water uses that have a significant impact on coastal waters;

- implementation of additional management measures, where necessary, to meet or protect water quality standards and to protect the waters of critical coastal areas;

- use of enforceable policies and mechanisms to implement the management measures; and

- program coordination to ensure consistency of this new coastal zone program with Clean Water Act programs under sections 208, 303, 319, and 320.

For those who had grown weary of the haphazard nature of the BMP lists in the state runoff control programs under section 319, the CZARA program looked like it might provide fairly seamless coverage of water-sensitive practices across wide swaths of coastal zones. The second major advantage of CZARA over the 319 program is that it requires EPA to provide the states with definite guidelines for water-sensitive practices. Under CZARA, states will have to implement management measures in their coastal zones that are consistent with EPA's minimum standards, thus removing some of the randomness (and weakness) that characterizes many 319 programs. The agency issued its final CZARA management guidance in January 1993.[60]

Environmentalists and some progressive state administrators urged EPA and NOAA (who jointly administer the program) to base management measures on objective, measurable criteria to ensure their effectiveness and accountability from state to state. Unfortunately, EPA and NOAA did not always heed this advice. For example, the draft guidance for controlling sediment pollution from farms originally would have required farmers to reduce erosion to specified levels (the T soil loss tolerance standard). The Natural Resources Defense Council and several other organizations supported this standard. Although less than perfect, it would afford an objective performance standard around which each coastal zone farmer could structure site-tailored erosion controls. In the final guidance, however, EPA caved to pressure from commodities groups and other agricultural special interests and recast the agricultural erosion control measure as the Alternative Conservation Systems (ACSs) described in the field office technical guides of the SCS. These ACSs are generally sound practices, but they provide little objective guidance for judging whether a farm has adopted sufficient erosion control practices.

Despite this weakening of the performance requirements for some management measures, however, the CZARA program remains a model for strong state runoff reduction programs. State implementation of required management measures for each land use category would improve 319 programs greatly if it were adopted for all watersheds, not just those in the coastal zone.

The coastal zone runoff program also contains some management measures, like vegetated riparian buffers, designed to protect and restore urban waters. A multitude of runoff sources severely degrade urban watersheds. Federal and state money and leadership are needed to create community programs that restore urban streams to full vitality; we describe these problems and solutions in the next section.

The Need for Urban Watershed Restoration

Urban waters are among the most degraded in the country. Urban streams are concretized and channelized, used as conduits for stormwater runoff, industrial and municipal effluents, and raw sewage from leaking sewer pipes (often laid lengthwise in streambeds) or from CSOs. And as if this were not enough abuse, many urban streams are obliterated altogether, "enclosed" (a euphemism for transforming a stream into an underground sewer) or, as in the case of many groundwater springs and first-order and ephemeral streams, simply destroyed beneath the treads of earth-moving vehicles preparing the ground for new development.

According to a 1992 EPA study of the environmental impacts of stormwater discharges, urbanization degrades a disproportionate share of our nation's waters:

> While urban population areas take up only about 2.5% of the total land surface of the country, stormwater pollution from these urban areas and associated urban activities (i.e., storm sewers/urban runoff, combined sewers, hydromodification, land disposal, construction, urban growth, etc.) accounts for a proportionately high degree of water quality impairment (i.e., 18% of impaired river miles, 34% of impaired lake acres, and 62% of impaired estuary square miles reported under 319) when compared to that from rural activities (i.e., agriculture, silviculture and mining) which take up approximately 53% of the total land surface.[61]

Urban stormwater pollution thus deserves high-priority attention by citizen activists, water quality officials, and other watershed stewards.

The most comprehensive study of urban runoff quality to date is the Nationwide Urban Runoff Program (NURP). This program, a joint project of

GS and EPA between 1979 and 1983, looked at stormwater quality in twenty-eight cities across the country. It found certain pollutants to be virtually ubiquitous in urban runoff, in average concentrations high enough to warrant concern over loadings in downstream sinks—estuaries like Chesapeake Bay and lakes like Quinsigamond in Worcester, Massachusetts. Among NURP's key findings:

● Copper, lead, and zinc were each found in at least 91 percent of the samples.
● Other frequently detected contaminants included arsenic, chromium, cadmium, nickel, and cyanide.
● Significant average concentrations of total suspended solids, phosphorus, nitrogen compounds, oxygen-robbing organic matter (BOD), and fecal coliform were found.[62]

Using national average runoff pollutant concentration data derived from the NURP study, NRDC made coarse estimates of runoff pollutant loadings for heavy metals, oil and grease, BOD, nitrogen, and phosphorus for seven urban areas around the country: Baltimore, Maryland; Washington, D.C.; Harrisburg, Pennsylvania; Tidewater, Virginia; Los Angeles, California; and Atlanta, Georgia. Although the results varied from city to city, these Poison Runoff Indexes showed that runoff rivals, and in some cases surpasses, factories and sewage plants as a source of these pollutants. For instance, in most of the urban areas modeled by NRDC, zinc loadings from runoff exceeded the loadings from factories.[63]

The NURP authors described the water quality impacts of urban runoff as falling into three categories:

1. short-term receiving water impacts during or following storm events, where pollutant concentration is important;
2. longer-term downstream receiving water effects: the buildup of contaminants in the sediments of sinks—river mouths, lakes, and bays—where seasonal or annual pollutant mass loads are important (although NURP did not examine in detail this phenomenon, NURP data enable coarse estimates to be made of runoff annual mass loadings from large urban areas); and
3. physical effects of storm flows on the hydrology and geomorphology of urbanized watersheds, including stream channel scouring (NURP did not examine this third type of effect but acknowledged its existence).[64]

One logical outcome of NURP's acknowledgement of this wide range in effects of urban runoff on receiving water is the creation of comprehensive watershed restoration programs. One example is the program developed for the Anacostia River, which flows through Washington, D.C., and into the Potomac River after collecting urban storm water from dozens of tributaries in suburban Maryland. The Anacostia is well known both for its severe degradation and for the extraordinary vision and commitment of the local governments now working toward its restoration. The six-point action plan for the Anacostia's restoration is keyed to a list of six problems that could apply to dozens of urban watersheds nationwide:

1. Poor water quality. The tidal Anacostia estuary has some of the poorest water quality recorded in the Chesapeake Bay system. It is rapidly filling with sediment and debris and has low dissolved oxygen levels and sediments contaminated with toxics.
2. Ecological degradation. Uncontrolled runoff and, in some cases, engineering "improvements" have degraded dozens of miles of stream habitat. Urbanization has profoundly altered the flow, shape, water quality, and ecology of these streams, many of which possess only a fraction of their original biodiversity.
3. Loss of anadromous fish habitat. As many as twenty-five human-made barriers prevent the upstream spawning migrations formerly made by menhaden, yellow perch, herring, and striped bass.
4. Loss of wetlands. More than 98 percent of the once-extensive tidal wetlands and nearly 75 percent of the watershed's freshwater wetlands have been destroyed.
5. Deforestation. Nearly 50 percent of the forest cover in the basin has been lost due to urbanization. The most severe losses have occurred in the riparian zones, where trees play a critical role in maintaining stream water quality, preventing streambank erosion, and providing both aquatic and terrestrial habitat.
6. Lack of public awareness. The six hundred thousand residents of the basin are generally unaware that they live in the Anacostia watershed. They do not perceive their connection to the river and its unique natural features or have the desire to take part in their watershed's restoration and become stewards.[65]

Despite all of this degradation, urban streams, lakes, and bays are still oases of life for millions of urbanites. Jamaica Bay is one example. Like many city water bodies, Jamaica Bay is oddly wild, given that it lies within the boundaries of New York City, it is bordered by Brooklyn and JFK Airport, and its waters are affected heavily by a mixture of urban runoff and sewage effluent. According to some of Jamaica Bay's stewards, "[F]ishing for sport and food has long been a favorite recreational activity in the park. Weekend fishermen line the railings of bridges and piers while others venture out in personal boats or charter fishing boats in hopes of a good catch."[66]

The City of New York Department of Parks and Recreation and the managers of the Gateway National Recreation Area recently surveyed 450 fishers who fish from the shores and bridges of Jamaica Bay. The survey revealed that 304 of the fishers, or two-thirds, eat the fish they catch, despite the fact that it is contaminated with low levels of PCBs.[67] And Jamaica Bay is not unique. People of all ages can be seen fishing for crayfish in Sligo Creek, an Anacostia tributary in Takoma Park, Maryland, and for catfish off of bridges over the Charles River outside of Boston. People fish regularly in Lake Erie off of the Fifty-fifth Street pier in Cleveland, Ohio, and off of wharves in South San Francisco Bay. The fact that at least some of these people eat what they catch, even if it may be contaminated, is not a reason to shut these active fisheries down. It is a reason to work with a sense of urgency to reduce and eliminate the toxics now flowing into them.

The 1987 amendments to the Clean Water Act included section 402(p), discussed in the previous section, which set forth stormwater permitting requirements for large and medium cities and for all industrial manufacturing sites that discharge storm water. Under EPA's implementation of 402(p), a total of 173 cities with populations of one hundred thousand or greater and forty-seven counties with unincorporated populations of one hundred thousand or more were required to have stormwater permits by October 1, 1992.[68]

Most of these municipalities have now applied for their initial permits (Part I) and have conducted stormwater pollution studies to develop citywide stormwater management programs (Part II). However, because EPA has not provided the states with substantive performance targets for the permits (such as the minimum urban area that must be covered by well-accepted stormwater management measures), urban citizens and stormwater utility ratepayers may have little or no assurance of permit program accountability and effectiveness. In addition, even EPA's own recent stormwater literature points out the need to expand the scope of regulation to additional large urban areas:

> The 220 Phase I NPDES municipalities have a combined urban population of 78 million. The remaining 80 million people located in urbanized areas are outside of Phase I municipalities. Most urban growth occurs in the urban fringe areas outside of core cities. For example, between 1970 and 1980, the population of incorporated cities with a population of 100,000 or more (Phase I cities) increased by only 0.6 million, with the population of many of these cities decreasing. Between 1970 and 1980, the population of urbanized areas outside of cities with a population of 100,000 or more increased 30 times more (an increase of 18.9 million) than the population of these core cities. This is important from a stormwater perspective as numerous studies (e.g., NURP) have shown that it is much more cost effective to develop measures to prevent or

reduce pollutants in stormwater during new development than it is to correct these problems later on.[69]

Thus, there are as many large urban areas currently outside of the NPDES stormwater permitting system as there are cities beneath the NPDES umbrella. This regulatory gap, as the quote from EPA above makes clear, is especially crucial considering that the areas left out of the NPDES umbrella are experiencing the most rapid growth rates and thus have the most urgent need for immediate establishment of water-sensitive master plans and site design practices before excavation and building ever begin. As one recent EPA report on the environmental quality impacts of land use observed, "[T]he significance of the [urban] sector is not how much land is in urban acres, but instead where the land is located, the implications and rapacity of recent development patterns, and the likelihood that future development will draw land out of other uses valued by society—agricultural lands, wetlands, or open space."[70] The concept of pollution prevention, a congressional mandate under the Pollution Prevention Act of 1990, ideally would work hand in hand with the Clean Water Act stormwater program under the following runoff prevention and reduction hierarchy:

1. for new development, runoff prevention through mapping and preservation of natural drainageways, preservation of mature forest zones along waterways, and caps on the amount of impervious surface;[71]
2. for redevelopment and retrofitting of existing developed areas, runoff reduction through revegetation and impervious surface reclamation (for example, retrofitting parking lots with grass swales designed to capture and filter the lot's runoff, thus preventing or severely reducing the need to discharge to a nearby stream);
3. chemical source controls and toxics use reduction (for example, implementation of policies that require lawn service companies to test lawns for nutrient content and pest problems before applying chemicals, in order to reduce lawn chemical use); and
4. conventional end-of-pipe stormwater treatment devices, such as extended detention ponds, infiltration trenches, and catch basins.

This stormwater policy hierarchy, in turn, could be incorporated into a comprehensive watershed restoration program that highlights the importance of urban waters to inner-city dwellers, relies on local citizen groups and municipalities to initiate and structure long-term restoration strategies (that may include community-based studies like surveys of urban fishing patterns and locally based skilled jobs like urban forestry), and channels federal dollars to priority urban watersheds to help fund the restoration work.[72] Such a program would help to focus the energies of urban activists on regreening the urban landscape, enshrining this ecology as a critical part of the Clean Water Act's goal of fishable, swimmable waters throughout the United States.

Conclusion

Federal and state water quality managers have historically missed out on opportunities to stem the flow of poison runoff via implementation of several key provisions of the Clean Water Act, most of which were available prior to the 1987 amendments. These key provisions include mandates to develop and apply relevant water quality standards; plan and manage whole watersheds; spread the burden of load reductions through TMDLs; and create focused, effective state runoff management programs. As a result of the failure to evolve these and other tools into effective runoff reduction and prevention programs on a watershed basis, poison runoff from virtually every category of land use continues to degrade the waters of the United States.

New federal and state programs, including the Coastal Zone Nonpoint Pollution Control Program and municipal and industrial stormwater permits, provide opportunities for states and EPA to eliminate the foot-dragging and unfocused, piecemeal approach to runoff control that occurred in the past. Whole-watershed management approaches are needed to tie together urban and rural dwellers in the goal of restoring their common waterways to full health. Such programs offer the promise that we can correct the mistakes of the past and actually stem the flow of poison runoff. Crucial to the success of these programs is the formidable political challenge of establishing enforceable requirements for water-sensitive land use practices and site designs that accrue to all of a watershed's landowners in a fair and equitable manner.

Chapter 7
Protection for Aquatic Resources and Ecosystems

Chapters 4 through 6 describe broken tools, each designed to fix particular insults to aquatic ecosystems. Each is needed in our overall efforts to protect water quality and aquatic resources, but taken together, these tools are not enough to address the range of problems discussed in chapter 2.

The Clean Water Act also includes provisions to protect specific aquatic resources. Section 303 and EPA rules require antidegradation programs to protect high quality waters, that is, to keep clean waters clean. Section 401 allows states to enforce water quality requirements against a wide range of federally licensed activities, such as dams and other water projects. Section 404 is designed to safeguard wetlands and other waters from the discharge of dredge or fill material or other major physical alterations. Other provisions protect particular types of ecosystems, such as lakes (section 314), estuaries (section 320), and oceans (section 403). Some programs focus on individual ecosystems, such as the Chesapeake Bay (section 117) and the Great Lakes (section 118), but other ecosystems of national or international significance, such as the Mississippi River, have not received equal treatment.

Each of these programs has been effective to some degree in protecting aquatic resources from pollution and other insults. However, as with the tools discussed previously, each also needs substantial improvement.

Moreover, each of the types of resources and ecosystems protected by these programs, such as wetlands, lakes, and estuaries, is part of a larger watershed. Each watershed may include wetlands, streams, rivers, lakes, ponds, estuaries, and coastal waters, or it may include some combination of these systems. A number of activities may impair each.

None of these tools used alone, therefore, will do the job. Each addresses discrete aspects of water pollution without a broad focus on the aquatic ecosystems on which we rely for drinking water, food, and other needs and amenities. To protect these systems using the available tools, we also need a set of plans and a coordinated effort to implement those plans. As implemented in many parts of the country, Clean Water Act efforts lack such a holistic effort.

Finally, in few cases have we devised comprehensive programs to restore previously impaired waters. We have restored some waters by eliminating the source of the problem—for example, by eliminating or reducing discharges from an existing point source. In time, the ecosystem may recover on its own. But little effort has been made to restore bodies of water currently suffering chemical or physical insults where natural restoration is impossible, where it will take long periods of time, or where acute human health and environmental threats remain from past pollution.

State Antidegradation Programs

States and EPA have given short shrift to antidegradation programs designed to keep clean waters clean. Few states have designated pristine or other highly valuable waters for special protection, and few state and federal agencies have implemented widespread programs to protect those waters that have been designated. As a result, some of the nation's most treasured aquatic resources, from our great national parks and wildlife refuges of Alaska to the Everglades in Florida, remain at serious risk.

State water quality standards programs have a key component that goes hand in hand with water quality criteria and use classifications: the antidegradation policy. Simply stated, the policy is intended to ensure that waters not yet degraded by pollution from point sources, polluted runoff, or other impacts remain that way. This policy springs from the key Clean Water Act mandate that we restore and maintain the physical, chemical, and biological integrity of our waters.

The antidegradation policy is critical in several respects. First, given the small percentage of waters that remain in a relatively pristine state, as discussed in chapter 2, we simply cannot afford to lose what we have left. Second, even where waters are partially degraded, it is wiser and cheaper in the long run to protect them from further degradation now than to allow them to follow the path of their impaired cousins and require remedial action later. Third, many of our relatively clean waters are in the headwaters of larger, more polluted watersheds. It makes little sense to clean up the main stems of larger rivers or downstream lakes and estuaries only to have them polluted from upstream waters. Like wetlands, headwater tributaries provide clean base flow and critical spawning and rearing habitat to support downstream ecosystems.[1]

In recognition of these needs, EPA issued regulations (last updated in 1983 but actually predating the 1972 Clean Water Act) calling on states to adopt and implement antidegradation policies to ensure that (1) we maintain and protect existing instream uses and the level of water quality needed to protect those uses; (2) we do not lower water quality where waters are cleaner than necessary to protect designated uses (that is, where waters are relatively clean) unless we first take all feasible steps to avoid doing so, and only in cases where lower water quality is necessary to accommodate important local needs; and (3) we maintain and protect water quality where high-quality waters constitute an outstanding national resource, such as waters of national and state parks and wildlife refuges and waters of exceptional recreational or ecological significance.[2]

This third antidegradation element, the Outstanding National Resource Waters (ONRW) provision, is intended to preserve the last, best, most unique aquatic environments from harm. The policy affords waters in this category the highest possible level of protection, because of their crucial importance to our ecological and aesthetic needs.

In a recent report analyzing the program's success in keeping clean waters clean, NWF evaluated the status of state ONRW programs to see whether we are, in fact, protecting our last, best waters. The results of their study were extraordinarily disheartening:

● Few states have adopted the federal program to protect outstanding waters, or developed equivalent state programs.

● As a result, only 0.4 percent of the nation's river miles currently are designated as ONRWs, and only another 3.8 percent of U.S. river miles are designated under similar state programs.

● Even where state programs do exist, federal land management agencies, with the exception of the National Park Service, have not used them effectively to protect outstanding waters either in or affecting important public lands.[3]

The survey found that only thirty-six states claimed legal authority to provide outstanding resource water protection. Of those thirty-six, only thirteen have any systematic method for identifying eligible water bodies, and only twenty-eight had actually designated one or more water bodies for protection.[4]

The National Wildlife Federation placed much of the blame for the fact that water bodies in our parks, wildlife refuges, and other ecologically important areas are at risk on EPA's shoulders because of its failure to provide the states with clear leadership and guidance. The federation took special issue with the agency's position that it does not have legal authority to review state decisions about whether or not to designate specific lakes, streams, or coastal waters as outstanding, even if the waters are located on public lands in national parks, wildlife refuges, forests, or wilderness areas. The agency also maintains that it lacks the legal authority to ensure that those state programs that do exist provide protection to outstanding waters equivalent to protection under the federal policy.[5] Without any serious threat of EPA action if they fail to comply, doubtless many states have found little incentive to act.

In addition to being lax in pressing states to protect key waterways, EPA has provided the other federal agencies that control vast stretches of federal lands, such as national parks and refuges, with little guidance on how to protect high-quality waters within their domain. It comes as no surprise, then, that with the limited exception of the National Park Service, federal agencies responsible for public lands have not acted systematically to protect waters in or affecting the lands entrusted to their stewardship. Sadly, NWF found that many federal land management officials, while acknowledging the importance of protecting high-quality waters and expressing great interest in enhancing water quality protection, were reticent about a strong federal initiative to designate and protect these outstanding waters.[6]

While no similar survey has been conducted to test the efficacy of state programs to protect other relatively clean waters (Tier 2 waters, which exceed water quality standards but are not appropriate for ONRW status), serious questions exist about this aspect of the antidegradation program as well. The most serious problem relates to the language of the EPA rule itself, which allows degradation where the state finds "that allowing lower water quality is necessary to accommodate important economic or social development in the area in which the waters are located."[7] The rule does not define "important economic or social development," thus providing states broad latitude for allowing degradation of water quality so long as the water quality standards are not exceeded. Another problem is that some antidegradation programs are limited in scope. Pennsylvania's program, for example, is limited to specially designated "High Quality" or "Exceptional Value" waters, in violation of the universal applicability of the EPA rule.

All told, our failure to establish and implement a robust program to prevent the degradation of our still-pristine and other relatively unimpaired waterways may turn out to be the biggest possible threat to the ultimate success of our national program to restore and maintain our waters. We don't need to lock up our still-unspoiled waterways and throw away the keys, but we do need to prevent the sprawl, paving over, plowing, and mining of the lands in and around our liquid jewels in order to preserve them for our children. Moreover, to achieve the ultimate goal of the Clean Water Act, we need to do more to prevent future degradation of all relatively clean waters, not just to take remedial action after the harm has occurred.

State Water Quality Certification

States have made relatively little use of their potentially powerful authority, through water quality certifications, to veto or condition federally licensed or permitted activities that may degrade water quality and aquatic habitat. Part of the problem lies in inconsistent federal and state court rulings on the scope of this program.

In chapter 5, we discussed problems in the implementation of water quality standards through the NPDES permitting program. Most analyses, however, ignore a second major means of implementing water quality standards, state water quality certification under section 401 of the act. This provision allows states to veto projects that may cause water quality problems, or to impose water-quality–based requirements on those projects. It applies to NPDES permits issued by EPA and to other federal licenses or permits to conduct "any activity," including hydroelectric projects, nuclear power plants, dredging and filling, and so on, "which may result in any discharge into the navigable waters."[8]

Section 401 potentially could be a potent tool for states to use to protect the integrity of their waters. It gives states water quality control over a wide range of activities for which they otherwise might lack such authority. Furthermore, in many respects, the provision is phrased in remarkably absolute terms: "No license or permit shall be granted until the certification required by this section has been obtained or has been waived."[9] And once a state issues a certification containing effluent limits or other standards, or "any other appropriate requirement of State law," such requirements "shall become a condition on any Federal license or permit subject to the provisions of this section."[10]

Some states have used section 401 to admirable effect. For example:

- Maine used section 401 to impose oil spill prevention as well as other water quality requirements on a proposed oil refinery and deep-water terminal.[11]
- Massachusetts denied water quality certification for a Corps' nationwide permit under section 404(e) of the act (discussed later in this chapter), requiring individual scrutiny for wetlands fills that otherwise would have been approved automatically.[12]
- The state of Washington conditioned its certification of a hydroelectric project on specific requirements designed to protect salmon.[13]
- Vermont exercised its authority to condition a 401 certification for a hydroelectric project to protect aesthetic and recreational values of the waterway.[14]
- Oregon was allowed to condition water quality certification on land use restrictions, so long as it could show that those restrictions were necessary to ensure compliance with water quality standards.[15]

Unfortunately, this authority is not used frequently, leading two commenters to refer to section 401 as the "Sleeping Giant."[16] Part of the problem may lie in the fact that states can waive water quality certification.[17] Moreover, section 401 on its face applies only to federally licensed or permitted activities, and confusion exists over which federal activities are actually subject to the requirement.[18] States strapped for resources may ignore, or give only short attention to, the myriad activities, aside from actual point source discharges, that may impair water quality or aquatic resources.

The authors of an article based on a 1987 EPA workshop on 401 certification found that "states are reviewing very few types of permits for water quality certification."[19] One telling example, given the findings in chapter 2 about the high percentage of waters impaired by physical alterations such as channelization, is that states rarely use water quality certification to review permits for such activities issued by the Corps under section 10 of the Rivers and Harbors Act.[20] And surveys conducted for this research revealed that many states rarely denied water quality certification even where permits were reviewed.[21]

Unfortunately, there is no central source of information on the workings of this potentially powerful section of the Clean Water Act. For example, EPA indicated that no systematic information is maintained on state 401 programs, and, worse yet, the issue is not even formally assigned to any single branch in EPA's Office of Water in Washington, D.C.[22] The only agency rules on state water quality certification were issued in 1971, based on the statutory predecessor to section 401, and "provide little guidance to state managers on substantive aspects of the program."[23]

Moreover, while some courts have interpreted section 401 with appropriate breadth (as summarized in the previously cited examples), others have narrowly constrained the activities states may review and the types of concerns they may address in water quality certifications under section 401. For example:

- A federal court in Pennsylvania ruled that section 401 certification was necessary only in the state in which the facility is located, not where the actual discharge is located.[24]
- A federal appeals court agreed in a case involving offshore oil and gas drilling that states receiving the impacts of federally licensed activities in federal waters (in this case Florida) have no authority to deny or condition water quality certification.[25]
- A state court in Pennsylvania found that the state could not base its water quality certification on physical and biological impacts (such as impacts to wetlands and fish migration) but was limited to chemical changes related to direct discharges of pollutants.[26]

● Courts in New York state repeatedly have limited certification authority to violations of chemical water quality standards.[27]

These and other cases pose several serious problems. As shown in chapter 2, many of the most serious impacts to aquatic resources are physical and biological, not purely chemical in nature. States are just beginning to address these broader impacts in their water quality standards and monitoring programs and have rarely done so in their water quality certifications. If these courts are correct that states may use section 401 only to address chemical impairment, how can states protect the physical and biological integrity of their waters, as required by the act? If fisheries and other aquatic resources are being destroyed by physical rather than chemical impacts, is it not artificial to limit states to tools that address only chemical pollution? Since the courts are divided on this issue, the overall success of the 401 program may turn on congressional clarification of state authority in this area.

Efforts to Protect Waters from Dredging and Filling

Section 404 of the Clean Water Act protects wetlands and other waters from the discharge of dredge and fill. But there are serious problems with implementation of this program. The vast majority of activities that cause damage to wetlands are exempt from the program; the definition of wetlands is shrouded in controversy; few permits to destroy wetlands are ever denied, and the cumulative impacts of these losses are poorly understood; wetlands restoration is difficult and unproven; and permits issued under the program are poorly enforced.

While section 402 of the Clean Water Act (the NPDES program discussed in chapter 4) regulates direct discharges of wastewaters containing pollutants, section 404 of the law authorizes the Corps to issue permits "for the discharge of *dredged or fill material* into the navigable waters at specified disposal sites."[28] Such permits are to be issued under EPA guidelines[29] and may be vetoed when EPA determines that the discharge would have "an unacceptable adverse effect on municipal water supplies, shellfish beds and fishery areas (including spawning and breeding areas), wildlife, or recreational areas."[30]

Specifically exempted from the permitting requirement, however, are activities such as normal farming, ranching, and logging; construction or maintenance of farm or stock ponds or irrigation ditches; maintenance (but not construction) of structures such as dams, levees, and bridges; use of temporary sedimentation basins at construction sites; and construction of farm or forest roads.[31] Moreover, under section 404, the Corps can regulate whole categories of activities through general permits issued on a state, regional, or nationwide basis. The statute allows such permits only where the activities "are similar in nature, will cause only minimal adverse environmental effects when performed separately, and will have only minimal cumulative adverse effect on the environment."[32]

While section 404 is associated most frequently with wetlands protection, this provision actually governs the discharge of dredged or fill material into all waterways regulated under the act.[33] Regulation of wetlands remains the most contentious focus of national debate, but virtually all of the regulatory issues surrounding wetlands also apply to discharges into other waters.

In recent years, section 404 has been the subject of the most scrutiny and the most intense opposition of any Clean Water Act provision. Opponents argue, for example, that the program unfairly restricts farming and other activities on private property; that less attention should be paid to "low-value" wetlands; and that the regulatory program is unduly slow and cumbersome, forcing innocent landowners to wait unreasonable amounts of time before they can develop their property.

The record, however, leads to the opposite conclusion. As the discussion of wetlands in chapter 2 indicated, we continue to lose almost 300,000 acres of wetlands each year, despite the regulatory program under section 404. While the rate of wetlands loss has slowed since passage of the Clean Water Act, it continues at an unacceptable rate. We demonstrated in chapter 2 that other damaging physical alterations of the nation's waterways, such as channelization and the placement of other physical obstructions, also continue at a rapid pace. This section discusses why these problems continue.

Inadequate Coverage of the 404 Program

Currently, section 404 applies only to the discharge of dredge or fill material. Other activities that destroy wetlands, such as draining, water diversion, excavation, clearing, channelizing and flooding, and placement of pilings or other obstructions, are not (or at least are not clearly)[34] within the purview of the permitting program. Moreover, as explained previously, the statute specifically exempts normal farming activities and other common activities from the permitting program.

Because of these gaps, the congressional OTA and the GAO have found that the Clean Water Act does not even govern the vast majority of ongoing wetlands losses, mostly because they resulted from agricultural activities.[35] This trend, however, is shifting somewhat. Whereas 87 percent of all wetlands losses from the mid-1950s to the mid-1970s were due to agriculture, this percentage dropped to 54 from the mid-1970s to mid-1980s. Most other losses during this period (41 percent) were due to draining and clearing of wetlands that were not yet put to an identifiable use, while 5 percent were due to urban development.[36] Still, a 1991 GAO study estimated that section 404 covers only one-fifth of the activities that destroy wetlands.[37]

This does not mean that the existing program is completely ineffective in protecting wetlands and other waters. For example, OTA estimated in 1984 that the existing regulatory program reduced wetlands losses by about 50,000 acres each year, roughly half of the acreage for which development was proposed.[38] And in recent years, Congress has sought to slow the loss of wetlands to agricultural activities through incentives programs under the 1985 and 1990 Farm Bills. The fact remains, however, that most actions that destroy the wetlands we have left are still outside the purview of the act.

Confusion Over Which Wetlands Are Covered

In addition to the issue of *what activities* are regulated by section 404, signifi cant controversy has arisen over the question of *which waters* are covered by the program. Section 404 clearly applies to such waters as rivers, lakes, bays, and estuaries. More controversy, however, surrounds the extent to which section 404 regulates activities in wetlands. This question has been the subject of considerable discussion elsewhere, including in an extensive and informative report by the EDF and the World Wildlife Fund (WWF).[39]

Early in the history of the Clean Water Act, some argued that section 404 did not apply to wetlands at all, but the courts rejected this argument.[40] In fact, in 1985, the U.S. Supreme Court rejected a broad statutory and constitutional challenge to the application of section 404 to wetlands, at least with respect to wetlands that are directly associated with other surface waters.[41]

Because wetlands span the transition zone between land and water,[42] however, a much harder question is, What is a wetland for purposes of Clean Water Act coverage? This already-complex scientific question is complicated further by the fact that multiple federal and state agencies delineate wetlands for different statutory purposes, sometimes with differing results. For purposes of the Clean Water Act, however, EPA and the Corps use a common regulatory definition of wetlands: "[T]hose areas that are inundated or saturated by surface or groundwater at a frequency and duration sufficient to support, and that under normal circumstances do support, a prevalence of vegetation typically adapted for life in saturated soil conditions. Wetlands generally include swamps, marshes, bogs, and similar areas."[43]

While this is the legally binding definition of wetlands, it provides only limited guidance to regulators trying to sort out the precise boundary between wetlands and uplands in a wide array of circumstances and different types of wetlands around the country. To provide more detailed direction to agency officials working in the field, a series of federal documents has been issued to help in wetlands delineation, beginning with the FWS's 1979 *Classification of Wetlands and Deepwater Habitats of the United States.*[44] One of the leading scientific texts on wetlands referred to the definition in this document as "the most widely accepted by wetlands scientists in the United States today":[45]

> Wetlands are lands transitional between terrestrial and aquatic systems where the water table is usually at or near the surface or the land is covered by shallow water.... Wetlands must have one or more of the following three attributes: (1) at least periodically, the land supports predominantly hydrophytes [plants that dwell in waterlogged soils], (2) the substrate is predominantly undrained hydric soil [containing water in the root zone long enough to eliminate oxygen], and (3) the substrate is nonsoil and is saturated with water or covered by shallow water at some time during the growing season of each year.[46]

This three-part test—plants, soils, and hydrology characteristic of wetlands—was incorporated into a series of federal wetlands delineations manuals, including one issued by the Corps in 1987, one by EPA in 1988, and one by the SCS in 1988.[47] In 1989, in an effort to eliminate confusion and provide uniform federal criteria for delineating wetlands, all four federal agencies (FWS, the Corps, EPA, and SCS) released a joint *Federal Manual for Identifying and Delineating Jurisdictional Wetlands.*

Many hailed the 1989 manual as a much-improved, if still not perfect, tool for identifying wetlands.[48] Like its predecessor manuals, the 1989 version identified the degree to which the three criteria had to be met and the types of evidence that would suffice to support a positive wetlands determination (the details of which are discussed in the EDF–WWF study). Unfortunately, much confusion resulted from misinformation about the 1989 manual and its implications. For example, many farmers were led to believe that large portions of their land previously exempt from regulation under section 404 now would be covered, simply because they appeared on maps labeled as having hydric soils. This was wrong for at least three reasons: (1) these maps were developed using a broad geographic scale sufficient to identify general regions but not to identify soils in a specific field; (2) hydric soils alone do not support a wetlands delineation without evidence of other factors; and (3) many of the ongoing farming activities on these lands were exempt from the 404 program even if they occurred in wetlands.

Confusion over the 1989 manual led to unfortunate backlash. The manual was challenged unsuccessfully in court.[49] Undaunted, opponents of the new manual turned to Congress, where they succeeded through an obscure rider to an appropriations bill in preventing the Corps from using the 1989 manual. (Recently, for purposes of consistency, EPA agreed to use the 1987 manual as well until another revised version is completed.) Misinformation about the manual and the program in general also led to the introduction in Congress of bills that would weaken substantially the entire section 404 program.[50]

Finally, political pressure to weaken wetlands regulation led the administration to hijack efforts by the four involved federal agencies to fine-tune the 1989 manual, and to release instead a proposed new manual that would reduce substantially the scope of the section 404 program. A detailed evaluation by EDF and WWF, with the help of professional wetlands scientists around the country, estimated that the new proposed criteria would eliminate coverage for half of the nation's remaining wetlands, and even more in some parts of the country. According to this report, for example, the new criteria would exclude almost one-quarter of Everglades National Park and more than 40 percent of the Everglades in private ownership; 80 percent of the Great Dismal Swamp in Virginia and North Carolina; more than one-third of the prairie potholes of North Dakota, which are crucial to migratory waterfowl populations; most of the bogs in the Northeast and Midwest; half of the hardwood swamps in the Southeast; and most of the high coastal marsh along the Pacific coast.[51]

Critics of the proposed new manual included not only environmental groups, but also a wide range of professional wetlands scientists. For example, the ad hoc Committee on Wetlands Delineation of the Ecological Society of America stated:

> We do not find the proposed revisions...to be a responsible application of ecological data or principles to the problem of wetland identification and delineation. In fact, no data or documentation are presented for the 1991 revisions. No citations to relevant recent studies or results of various tests of previous versions of the Manual are provided. Well-established scientific understanding of wetlands appears to have been ignored. The 1991 revisions lack the scientific justification that ought to be required for what is considered "a technical guidance document."[52]

More succinctly, the chief ecologist in EPA's wetlands division wrote, "In my opinion, the 1991 revisions are technically flawed, operationally useless, and economically impractical."[53]

Meanwhile, as part of the 1993 EPA appropriations bill, Congress directed NAS to prepare a study of key disputed questions about wetlands delineation, in an effort to bring science rather than politics back into the equation. Although the results of this study technically are due in November 1993, the final report is not expected until mid-1994 or later.

The Bush administration ultimately abandoned its efforts to issue its new federal delineation manual, and wetlands delineation by the Corps and EPA for purposes of 404 program coverage are being performed under the 1987 Corps manual. Until the issue is resolved, however, either by returning to the 1989 manual or by adopting an improved version based on the results of the NAS study, a cloud will remain over implementation of the entire section 404 program.

Failure to Prohibit or Mitigate Properly under Section 404

Despite complaints that section 404 unduly restricts activities on private property, most proposed activities are allowed under it, either through individual, general, or nationwide permits. Of approximately 15,000 project-specific permit applications evaluated by the Corps each year, about 10,000 are issued, 500 are denied, and 4,500 either qualify for coverage under a general permit (and therefore are allowed) or are withdrawn or canceled. Another 40,000 activities occur each year under regional and nationwide general permits.[54] In sum, between 50,000 and 54,500 of the approximately 55,000 activities proposed under section 404 actually occurred each year between 1988 and 1990.

General Permits

A major portion of this activity occurs under the auspices of general permits. Currently, there are thirty-six nationwide permits covering a wide range of activities,[55] as well as regional general permits issued by individual Corps district offices. For twenty-five of these thirty-six types of permits, the landowner does not even need to inform the Corps of the activity before proceeding.[56]

The broadest and most controversial of these general permits is Nationwide Permit 26. This permit exempts from individual requirements all activities that affect less than 1 acre in "isolated wetlands"[57] or that are in areas above the headwaters of nontidal rivers and streams. For projects that affect between 1 and 10 acres in these areas, notification is required, but approval is granted automatically if not denied within thirty days.

Unfortunately, little information is available on the acreage of wetlands and other waters destroyed or altered because of general permits. In fact, a GAO survey of several Corps offices revealed that virtually no comprehensive information is maintained on total wetlands losses from general or individual permits.[58] Thus, it is impossible to judge the overall impact of general permits on aquatic resources.

Two FWS biologists suggested the serious dangers inherent in the general permits program in a study on the effect of Nationwide Permit 26 in the Platte

River basin in Colorado and in the Sacramento River valley in California. The study concluded that activities were poorly controlled under Nationwide Permit 26 because

● the lack of opportunity to comment on projects of less than 1 acre eliminated the potential to evaluate effects on threatened and endangered species;

● mitigation cannot be requested for projects of less than 1 acre, so wetland restoration for such projects was virtually nonexistent;

● projects conducted under the permit received little or no scrutiny for compliance with program requirements, including those that might prohibit the action; and

● discharges authorized by the permit "composed a significant fraction of the total discharges permitted by section 404 in both northeastern Colorado and north-central Colorado."[59]

The authors expressed concern that the cumulative effects of activities under Nationwide Permit 26 may be significant in Colorado, where "wetlands compose less than five percent of the land area but are used by 90 percent of the wildlife species," and in California, "where over 90 percent of the state's historic wetlands have already been destroyed."[60]

Practicable Alternatives and Water Dependency

Even where individual permits are required, the Corps approves the majority of proposed activities. Ironically, EPA's guidelines for the issuance of section 404 permits impose some relatively stringent review requirements that should result in the disapproval of many activities, but it appears that these strictures are applied more as the exception than as the rule.

The section 404 regulatory program is far more complex than can be recited fully here, but the two major components of EPA's guidelines are the "practicable alternatives" and "water dependency" tests. Taken together, these rules establish a legal presumption against issuance of permits in "special aquatic sites," which include wetlands, sanctuaries, refuges, mud flats, and vegetated shallows.[61] First, for all activities, the guidelines establish a presumption against filling "if there is a practicable alternative to the proposed discharge which would have less adverse impact on the aquatic ecosystem."[62] The presumption becomes even firmer for proposed activities that are not water dependent. (Water-dependent activities include, for example, marinas or other activities that obviously must be adjacent to the water.) For non-water-dependent activities, "practicable alternatives that do not involve special aquatic sites are presumed to be available, unless clearly demonstrated otherwise."[63]

If the applicant bears such a heavy burden to overcome these presumptions, why are such a high percentage of permits issued under section 404? In 1988, GAO explained that Corps officials frequently accepted whatever statement the applicant provided about the purposes of and need for the project, and hence about the absence of any practicable alternatives.[64] The GAO report quoted an EPA official, who wrote, "In our experience for the majority of cases we have seen, the Corps practice is to issue permits for whatever the applicant wants with very little consideration given to the 'tests' within the Guidelines that address prohibition and alternatives, or EPA stated concerns."[65] Subsequently, the Corps revised its regulations to indicate that it would apply more independent scrutiny to the applicant's statements of project need.[66] Little empirical or qualitative information appears to be available, however, on whether the Corps indeed has applied a new approach and on whether more scrutiny is being provided under the practicable alternatives and water-dependency tests.

Doubts about Mitigation and Restoration in Lieu of Protection

Environmental Protection Agency regulations also prohibit the issuance of 404 permits "unless appropriate and practicable steps have been taken which will minimize potential adverse effects of the discharge on the aquatic ecosystem."[67] The agency and the Corps have disputed the meaning of this regulation. The Corps argued that mitigation could be used to satisfy other requirements of the regulations. But EPA believed that mitigation was only acceptable as a last resort, that is, after demonstration that harm could not be avoided completely, minimized, or restored after any damage from the project occurred.[68]

The Corps and EPA resolved this dispute in a Memorandum of Agreement on Mitigation, which, after several fits and starts, went into effect in February 1990. This document largely adopted EPA's sequencing approach, under which "the Corps...first makes a determination that potential impacts have been avoided to the maximum extent practicable; remaining unavoidable impacts will then be mitigated to the extent appropriate and practicable by requiring steps to minimize impacts and, finally, compensate for aquatic resource values."[69] In short: avoid first, minimize second, restore and compensate as a last resort.

The mitigation controversy harkens back to the findings of the National Wetlands Policy Forum, a policy dialogue convened by the Conservation Foundation in 1988 with the goal of bringing together the various interest groups with a stake in the wetlands debate. Among a wide range of specific recommendations on wetlands protection, the most enduring advice in the forum's final report was that "the nation establish a national wetlands protection policy to achieve no overall net loss of the nation's remaining wetlands base, as defined by acreage and function, and to restore and create wetlands, where feasible, to increase the quality and quantity of the nation's wetlands resource base."[70] "No overall net loss" was specified as an interim goal, with an increase in the "quantity and quality of the nation's wetlands resource base" articulated as the long-term goal.[71]

Then-Vice-President George Bush endorsed the interim goal of no net loss in the 1988 presidential campaign, and the National Governor's Association adopted it in 1988 and 1991.[72] Accordingly, "no net loss" was enshrined in the 1990 Memorandum of Agreement. Lost in the shuffle, however, has been the second component of the Wetlands Policy Forum's statement, the long-term goal to increase the nation's wetlands resource. At best, the goal of no net loss comports only with part of the Clean Water Act's most fundamental command, to "restore and maintain the...integrity of the Nation's waters." A policy that ignores the long-term goal of increasing wetlands resources fails even to give lip service to restoration, a crucial goal given the loss of half of our historical wetlands resources.

More important from a practical if not a philosophical perspective, however, is the viability of mitigation and restoration in the first place. If we continue to allow hundreds of thousands of acres of wetlands to be filled and paved each year, how can we achieve even no net loss, much less an increase in wetlands values and functions? Unless we reduce dramatically any future wetlands losses, the answer can lie only in the feasibility of wetlands restoration and creation.

This issue is the subject of tremendous scientific debate. But while some claim that wetlands can be restored or "created," the body of scientific evidence suggests that wetlands restoration is an unproven science. According to the NRC:

> Mitigation efforts cannot yet claim to have duplicated lost wetland functional values. It has not been shown that restored wetlands maintain regional biodiversity and recreate functional ecosystems. There is some evidence that created wetlands can look like natural ones; there are few data to show that they behave like natural ones. In many cases, scientific knowledge of how to restore degraded wetland systems is limited, and wetland creation has been largely a matter of trial and error.[73]

Moreover, the council notes that most partially successful wetlands restoration efforts have been in coastal wetlands. While these restoration efforts have not been fully documented, far less effort has been made to restore inland wetlands, including headwater and depressional wetlands (such as prairie potholes), which may be even more difficult.[74]

Thus, from the perspective of compensating for ongoing damage to wetlands and other aquatic ecosystems, the use of mitigation and restoration should be viewed with extreme caution. Nevertheless, given the tremendous losses already incurred, clearly we must honor the restoration goal of the Clean Water Act through restoration programs independent of future permitting decisions. In this vein, NRC recommended a goal of restoring 10 million acres of wetlands by the year 2010 (above what is lost through future activities during that period).[75] This represents less than one-tenth of what has been lost over the past two centuries. Because most losses have been freshwater wetlands in agricultural regions, the council recommends that any national restoration effort focus on these areas.[76]

Failure to Consider Cumulative Impacts

Both EPA and Corps regulations require the agencies to consider the cumulative as well as the individual impacts of section 404 permits.[77] The NRC noted that "the cumulative impact of many individual actions, no single one of which is particularly alarming, threatens the integrity of entire wetland landscapes."[78]

Unfortunately, cumulative impacts are rarely considered in the 404 permit review process. Indeed, agencies have yet even to develop an accepted methodology for cumulative impacts assessment.[79] The lack of cumulative impacts analysis is particularly problematic for general permits.[80]

Agency Recommendations Often Ignored

In theory, the Clean Water Act and other laws give EPA and FWS particular influence over the Corps' permitting decisions. The Fish and Wildlife Service is instructed to comment on all individual and general permits issued by the Corps. The Environmental Protection Agency can play an even greater role through its authority to veto permits under section 404(c). Other federal, state, and local agencies, as well as the general public, may comment on proposed permits as well. Unfortunately, the Corps frequently ignores these agency comments. For its part, EPA has used its veto authority only in the rarest of cases.

A 1988 GAO survey of several Corps districts found that the Corps actually issued permits over the objection of another federal agency in more than one-third of the cases in which permit denial was recommended. While the Corps accepted more agency suggestions that fell short of outright permit denial, it rejected even these proposals in one out of five cases.[81] Moreover, because agency officials believe their ability to overturn these decisions through elevation is limited and time-consuming, very few interagency disputes are appealed. From 1982 to 1986, for example, no decisions were elevated in two Corps districts (Baltimore and Omaha), and only one was appealed in the Portland region. In five Corps districts combined, only fifty-two permit decisions were elevated over the five-year period.[82]

The Environmental Protection Agency has vetoed Corps permits under section 404(c) even less frequently. As of 1988, GAO reported only five EPA vetoes in the sixteen-year history of section 404.[83] Since that time, EPA has vetoed at least two other permits, both highly public and laudable (the Two Forks dam project in Colorado and the Ware Creek project in Virginia). Still, given the thousands of Corps permits issued each year, this record hardly demonstrates serious EPA scrutiny on a nationwide basis.

Weak Enforcement of Section 404

Although there is evidence of improvement in recent years, neither EPA nor the Corps has acted aggressively to enforce section 404, against either unpermitted activities or actions taken in violation of permits. The 1988 GAO report found the following:

● The Corps rarely performed surveillance to detect unpermitted activities or to ensure compliance with permit conditions. The Jacksonville District, for example, spent only about 5 percent of its time on surveillance, and other district offices agreed that enforcement efforts were lacking. Compliance was verified for only a fraction of all permits.

● Even after reports of illegal activities were received, some investigations were delayed for months or not conducted at all. While some investigations occurred within five days of notice, many were delayed for one to three weeks, and some were delayed for months.

● Enforcement actions that did occur relied primarily on administrative, rather than civil or criminal, remedies. From 1984 to 1986, for example, only six civil and no criminal cases were reported in the five districts surveyed, and only two fines were imposed for program violations.

● Few permits were revoked or suspended due to noncompliance with permit terms. From 1984 to 1986, no revocations and only sixteen suspensions were identified.[84]

The Environmental Protection Agency has independent authority to enforce the section 404 program. The General Accounting Office concluded that EPA had used this power sparingly, although the statistics provide preliminary evidence that EPA had taken a somewhat more aggressive enforcement posture than had the Corps. During 1986 and 1987, EPA took 194 actions nationwide to enforce the 404 program, although there was considerable variation among regions. During this two-year period, EPA Region VI took only one action and Region VIII only six, with the majority of cases brought by Regions III and IV.[85]

Programs to Protect Lakes, Estuaries, Ocean Waters, and Rivers

The Clean Water Act also includes programs to address specific types of water bodies. These include the Clean Lakes Program (section 314), the National Estuary Program (section 320), Ocean Discharge Criteria (section 403), and the Chesapeake Bay (section 117) and Great Lakes (section 118) programs. While some individual rivers are subject to interstate compacts and cooperative management under other statutes (such as the Ohio River Basin Sanitary Commission and the Delaware River Basin Commission), there is no comparable national rivers program.

Each of these programs has resulted in some progress beyond that achieved through the individual control provisions discussed in the previous sections of this chapter. All suffer, however, from common defects: (1) They are not complete—that is, they do not cover the full range of impaired waters and the full range of problems affecting those waters; and (2) they are inadequately funded. Moreover, these programs treat each type of aquatic system individually, rather than as part of an integrally linked system, and fail to address comprehensive restoration efforts of all types of waters within whole watersheds. And while some of these programs incorporate restoration as well as prevention of future harm, none creates the comprehensive restoration effort necessary to reverse more than two centuries of severe physical, chemical, and biological damage to aquatic ecosystems in the United States.

The Clean Lakes Program

Congress provided a small degree of extra consideration to lakes in section 314, the Clean Lakes Program. Under this provision, states can get grants to assess lake quality, including eutrophication, acidity, and other problems common to lakes, as well as methods to restore lake water and habitat quality. State 305(b) reports and EPA's biennial reports to Congress include these assessments.

In its 1989 report to Congress on lake water quality, EPA noted that one-quarter of the more than 12 million lake acres assessed were impaired, and pollution threatened another 20 percent. Not surprisingly, the dominant pollutants nationally were nutrients and siltation/turbidity, with the overwhelming majority of these pollutants coming from agricultural and urban runoff. While little information was available to assess trends accurately, EPA found a 10 percent rise in the number of lakes considered eutrophic over the previous two years.[86] Other information confirms that, in general, the quality of the nation's lakes is deteriorating in many respects.[87] And even these data are subject to the monitoring gaps and problems discussed previously.

Section 314 also established a demonstration program for lake restoration techniques. From 1975 to 1985, funds were provided for restoration projects on 313 lakes in forty-seven states and Puerto Rico.[88] Taken as a whole, however, this program is at best a small bite out of a large apple.

For one thing, there are thousands, not hundreds, of impaired lakes around the country. The NRC has described the 1991 budget for the Clean Lakes Program ($8 million) as "minuscule relative to the large task of restoration facing the United States."[89] The council believed an attainable goal would be to restore 1 million of the approximately 4.3 million acres of degraded lakes in the United States by the turn of the century, and another million acres (for a total of 2 million) by the year 2010.[90]

Moreover, the existing skeleton program hardly qualifies as a comprehensive lake management and restoration program. Only the vaguest guidance is provided as to how program funds are to be spent, with no minimum requirements for comprehensive monitoring, identification of sources of impairment, and comprehensive remediation. As noted in the NRC report, lake restoration "has strong interactions with restoration of other watershed components," including influent rivers and streams and surrounding wetlands.[91] In other words, restoring lakes themselves may be a futile effort if adjacent wetlands are destroyed (or left impaired) or if pollution from upstream sources continues. And prevention of future pollution may not reap immediate benefits if toxics and nutrients in sediment remain in place. The Clean Lakes Program, as currently designed, lacks both funding and a program structure to implement a comprehensive, watershed-based lake restoration program on a national scale.

The National Estuary Program

Considerable evidence indicates that water quality in many of the nation's estuaries is degrading. In a comprehensive 1987 report, OTA concluded that the overall health of estuaries is "declining or threatened" due to a wide range of human impacts and that even compliance with all existing requirements would not suffice to restore the health of all of these critical coastal ecosystems.[92] Estuaries receive the vast majority of all pollution discharges to coastal waters, OTA concluded.

The National Estuary Program (NEP), added to the Clean Water Act (as section 320) in 1987, comes closer than the Clean Lakes Program to being a comprehensive watershed-based program. The program allows the governor of any state to nominate an "estuary of national significance." If EPA finds that supplemental controls are necessary to attain the water quality goals of the act, these estuaries are eligible for funding to convene a conference to develop a comprehensive management plan. Congress initially designated eleven estuaries for priority consideration, and the program has now grown to include seventeen around the country.[93]

Management conferences are intended to assess trends in water quality and natural resources; evaluate pollution and other sources of impairment; and develop and implement comprehensive corrective actions to restore and maintain the chemical, physical, and biological integrity of the estuary. Designed as a participatory process, NEP management conferences are to include representatives of affected states and foreign nations; federal agencies; local governments; affected industries and educational institutions; and the public.

The NEP strives to be a true comprehensive watershed initiative, instructing that all sources of impairment be assessed and that remedial actions be designed and implemented to tackle these problems. Like the Lakes Program, however, NEP suffers in scope and ultimate commitment to implementation. The seventeen estuaries now included in the program, for example, include just 11,621 out of the estimated 35,624 square miles of estuaries in the country, less than 1 out of 3 square miles.[94] Still, this covers a far greater percentage of the nation's estuaries than the Clean Lakes Program does for lakes.

More important, while section 320(f)(2) of the Clean Water Act states clearly that "[u]pon approval...such plan shall be implemented," the law includes no firm way to ensure that the public will get appropriate action for its money. First, while the provision identifies items to be considered in plan development, few specific mandates limit EPA's ability to approve management programs as sufficient to meet the requirements of the act; and no provision requires that, as Congress intended, the program be used to go beyond the existing (non-NEP) requirements of the law. Second, future NEP grants are not conditioned on actual implementation of management plans. While no evidence indicates that the programs identified in the seventeen NEP plans will not be implemented, EPA reported in April 1992, more than five years after the program was created, that only two of the estuary conferences (Puget

Sound, Washington, and Buzzards Bay, Massachusetts) had completed management plans and moved fully into the implementation phase.[95]

Ocean Discharge Criteria

No equivalent comprehensive program exists to address ocean waters, which come under the jurisdiction of other federal statutes (in particular the Marine Protection, Research and Sanctuaries Act) as well as the Clean Water Act. At least one provision of the Clean Water Act, Ocean Discharge Criteria (section 403), in theory provides added protection to ocean waters, at least from point source discharges regulated under NPDES permits and discharges regulated under section 404.

Section 403 imposes additional requirements on discharges to ocean waters that do not apply to releases to other waters. In particular, EPA was required to judge such discharges against criteria designed to protect against adverse effects to fish, shellfish, wildlife, coastlines, beaches, and other economic and ecological values of marine resources.[96] Because of the broad array of values to be considered, EPA refers to the program as imposing whole ecosystem-based protection:

> EPA's section 403(c) program stresses consideration of the receiving water ecosystem, protection of unique, sensitive or ecologically critical species, and protection of human health and recreational uses. If technology-based and water quality-based limitations are...met by the discharger, but it is determined that the discharge still will cause an unreasonable degradation of the marine environment, then permit writers must impose additional restrictions on the discharge, including a prohibition of discharge if necessary...so that unreasonable degradation does not occur.[97]

Strongest of all, however, is section 403(c)'s flat statutory prohibition against the issuance of any permits "where insufficient information exists on any proposed discharge to make a reasonable judgment" on any of these impacts—in other words, if you don't know, don't act.

In 1990, EPA submitted to Congress a detailed report on the implementation of section 403. Based on the information in this report, substantial effort has been made to review marine NPDES permits for compliance with Ocean Discharge Criteria. The agency identified 323 individual NPDES permits and 10 general permits (9 for offshore oil and gas wells and 1 for seafood processors in Alaska) that have been subject to scrutiny under section 403; it indicates that the status of another 217 discharges, largely small POTWs in Alaska, is "uncertain."[98]

Notwithstanding the criteria in section 403(c), EPA's inventory reveals that NPDES permittees discharge massive amounts of pollutants into coastal waters. From individual dischargers alone, EPA estimates annual releases of more than .5 million tons of suspended solids and biological oxygen-demanding pollutants, more than 80,000 tons of nutrients (nitrogen and phosphorus), more than 50,000 tons of petroleum hydrocarbons, and almost 600 tons of toxic metals.[99] Offshore oil and gas operations operating mostly under general permits add at least 15 million barrels (more than 60 million gallons) a year of wastewaters from exploration wells alone (based on 1986 wells) and more than 8 million barrels (more than 40 million gallons) of toxic brines each day from production wells. The agency acknowledges that these wastes contain a wide range of toxic metals and organic chemicals, as well as large amounts of conventional pollutants.[100]

Most of what EPA characterizes as "large" dischargers are in compliance with Ocean Discharge Criteria, according to EPA. It is not clear, however, whether EPA believes that many small dischargers are *not* in compliance, or whether it simply does not know. Moreover, EPA confesses that there are, overall, substantial problems with the scope of 403(c) reviews: "As a result of the rapidly evolving nature of the permits program for marine waters and the limited availability of resources at the local and Regional levels, the detail of 403(c) reviews, the effectiveness of monitoring programs, and the level of review performed after permit issuance has varied by Region, by State, and by discharge."[101]

What EPA fails to highlight is the illegal fudge factor it slipped into the Ocean Discharge Program. As explained previously, the law prohibits any discharge where information is inadequate to make an affirmative finding that the criteria will be met. This reflected a clear congressional choice that, where scientific understanding of an ocean discharge is lacking, we should err on the side of protecting ocean waters and resources. But when EPA issued its ocean discharge rules to implement this program, it included a critical change. Rather than prohibiting discharges where information is inadequate to ensure that ocean resources will be protected, EPA allows such discharges so long as the resulting harm will not be "irreparable."[102]

In its 1990 report to Congress, EPA discusses its "irreparable harm" criteria but does not reveal that it changed Congress's will in such a significant way. It is difficult to judge how many of the thousands of tons of pollutants now being released into the ocean would be prohibited had EPA enforced the law faithfully. A hint is provided, however, by EPA's concessions about the types of tests and criteria currently excluded from the program, such as criteria to fully address sediment toxicity, aquatic toxicology, bioaccumulation, and biological integrity. In summary of the continuing problem, EPA notes that, even in the next round of 403(c) permits, "many determinations will be based on irreparable harm but that [in the following] phase the monitoring will have generated data to fill in the information gaps for assessing impacts using the ten 403(c) factors."[103]

Regional Ecosystems

The Clean Water Act contains special provisions to protect two major regional ecosystems on a cooperative basis. These are the Great Lakes (section 118) and the Chesapeake Bay (section 117). Both involve large aquatic ecosystems of interstate import; the Great Lakes also involve international import. Both provide models for interjurisdictional cooperation to tackle severe contamination and other impairment of aquatic resources from a range of human activities. And while many of the same threats and the same programmatic difficulties as discussed previously beset each ecosystem and each program, the Chesapeake Bay and the Great Lakes are also shining examples of progress in different areas. The Great Lakes Program is at the forefront of efforts to achieve zero discharge of toxic pollutants, while the Chesapeake Bay Program is, in many ways, leading the charge in our efforts to address poison runoff.

The Great Lakes and the Chesapeake Bay programs have been evaluated comprehensively in two recent books—*Turning the Tide, Saving the Chesapeake Bay*, by the CBF, and *Great Lakes, Great Legacy?* by the Conservation Foundation and the Institute for Research Public Policy[104]—and in other sources.[105] We will not try to duplicate these efforts here. Instead, we outline briefly the history and basic structure of each regional program; highlight the example that each provides for control of toxic pollution and poison runoff, respectively; and summarize key problems that have been identified in each effort.

Management of the Great Lakes enjoys a long history of international cooperation, dating to the establishment of the IJC in 1909. This cooperation led to the signing of a U.S.–Canadian Great Lakes Water Quality Agreement in 1972, followed by a more comprehensive agreement in 1978 (with further amendments in 1987). The 1972 agreement focused largely on the most imminent threat to the health of the Great Lakes—severe oxygen depletion due to excess releases of nutrients. As discussed in chapter 2, efforts to correct this problem have been successful but are still incomplete. The 1978 agreement adopted a broader goal that echoes the opening words of the Clean Water Act: "to restore and maintain the chemical, physical, and biological integrity of the Great Lakes Basin Ecosystem."[106]

United States implementation of the Great Lakes Water Quality Agreements is now conducted under section 118 of the Clean Water Act, which was added to the law in 1987 and amended in 1990. This provision establishes a Great Lakes National Program Office (section 118[b]), which has a range of authority and responsibility to promote implementation of the agreements by developing and implementing action plans, monitoring a surveillance, issuing water quality guidance, developing and implementing remedial action plans for acutely polluted areas, developing lakewide management plans, and performing other functions (section 118(c)).

In *Great Lakes, Great Legacy?* the Conservation Foundation and Canada's Institute for Research on Public Policy identify two notable elements of the Great Lakes agreement. The first is the agreement's whole-ecosystem approach to environmental management, which takes into account all sources of pollution (point source, runoff, and airborne) to all environmental media (air, surface, water, groundwater, and land) and their impacts on all inhabitants of the ecosystem (people, plants, and animals). Second, they highlight the commitment to "virtually eliminate" the release of persistent toxic pollutants into the Great Lakes ecosystem, with "persistent toxics" being "those that remain in the ecosystem for a long time."[107]

The commitment to zero discharge and virtual elimination remains the most persistent focus of the Great Lakes agreement. Yet while much progress has been made, the IJC recognized in 1992 that we are a long way from attaining these goals: "The Agreement calls for the virtual elimination of the input of persistent toxic substances into the Great Lakes basin to protect human and environmental health. We have not yet virtually eliminated, nor achieved zero discharge of any persistent toxic substance."[108]

To address this continued release of persistent toxic pollutants into the Great Lakes, the IJC proposed a bold, comprehensive program to ban or phase out (*sunset*) the release of toxic substances that remain in water, air, soil, or biota for more than eight weeks, and toxic chemicals that bioaccumulate in living organisms.[109] Moreover, the IJC recommended that Lake Superior be designated as a "demonstration area where no point source of any persistent toxic substance will be permitted" and recommended that a specific date be set after which the point source release of persistent toxic substances will be banned.[110] This proposal can serve as a model for a more comprehensive approach to achieving zero discharge in the Clean Water Act.

Environmental groups agree that improvements are needed in existing water quality programs in order to achieve the zero discharge goal of the Great Lakes Water Quality Agreements. Specific recommendations include banning or sunsetting the use and release of the most persistent toxic chemicals, establishing consistent or uniform fish advisory and water quality standards, developing sediment quality standards and a contaminated sediment inventory and cleanup program, and prohibiting the use of dilution and mixing zones for persistent, bioaccumulative, toxic pollutants.[111]

The Chesapeake Bay Commission and EPA's Chesapeake Interstate Bay Program office implement a second interjurisdictional water resource protection program that provides positive example for other cooperative regional programs. The Chesapeake Bay Commission was created in 1980 by the states of Maryland and Virginia and modified in 1985 to add the state of Pennsylvania. This joint effort led to the signing of interstate Chesapeake Bay agreements in 1983 and 1987.[112] As with the Great Lakes, Congress in 1987 added a specific provision to the Clean Water Act (section 117) to provide direction and funding for EPA's Chesapeake Bay Program office.

Like the Great Lakes Program, the Chesapeake Bay agreements provide an ecosystem-wide framework for protecting the rich natural resources of the Chesapeake Bay from a wide range of threats, including polluted runoff, declining fisheries, boat pollution, and oil spills.[113] The Chesapeake Bay cleanup is known best, however, for its broad efforts to address runoff pollution from land-based sources.

Two elements of the Chesapeake Bay program are noteworthy in this regard. First is the commitment in the 1987 Chesapeake Bay agreement to a 40 percent overall reduction in the release of phosphorus and nitrogen to the Bay from all sources. The commission's 1991 annual report indicated that the signatory states were well on their way to meeting this goal for phosphorus but that nitrogen inputs actually had increased since the 1987 Agreement had been signed. For both nutrients, the majority of remaining discharges (77 percent for nitrogen and 66 percent for phosphorus) are from agricultural, urban, and other sources of runoff.[114] The Chesapeake Bay Commission has not yet agreed on a comprehensive, enforceable program—like the IJC consensus on toxics for the Great Lakes—to ensure that the 40 percent reduction goal will be met. For example, the commission has not yet agreed whether the goal can be met through purely voluntary and "incentive-driven" programs, or whether mandatory approaches will be needed.[115] To some degree, Congress has preempted this issue with CZARA, discussed in chapter 6, which requires all coastal states to implement enforceable management measures for sources of polluted runoff.

The second notable element of the Chesapeake Bay Program is its recognition of the fundamental link between population growth and development and the health of the Bay. This connection led to formation of the Year 2020 Panel, which advised that growth management and sound land use planning were essential components of any program to protect the Bay. Implementation of these recommendations, however, was left to the individual states. This led to enactment of the Maryland Growth and Chesapeake Bay Protection Act and the creation in Virginia of a Commission on Population Growth and Development, but it is far too early to judge the effectiveness of these actions.[116]

Ecosystem Protection for Rivers Lacking

The NRC report on restoring aquatic ecosystems complains, perhaps justifiably, that "[a]t present, the collective federal water quality program emphasizes streams, rivers, and wetlands."[117] It is ironic, then, that while we have Clean Lakes and National Estuary programs, however deficient these efforts may be, there is no comparable national rivers program under the Clean Water Act.

Arguably, such programs were implied by the comprehensive watershed-based planning requirements that Congress intended to establish in sections 201 and 208 of the law but that, as explained in the section on poison runoff, were subverted by point source-dominated thinking and then deep-sixed in the budget battles of the early 1980s. Some comprehensive planning occurs for specific rivers, especially interstate rivers such as the Ohio, Delaware, and Colorado, under the auspices of river basin planning commissions established under separate federal laws. But other major river systems—such as the Mississippi (North America's most dominant watershed), the Columbia, and the Rio Grande—do not share such benefits; nor do hundreds of other watersheds around the country.

Certainly, the absence of a comprehensive program to protect and restore the nation's rivers is not due to lack of need. As shown in chapter 2, the vast majority of our river systems have been dammed, channelized, levied, paved over, or otherwise altered. Yet the NRC report notes that, while we at least have programs to deal with chemical pollution, "there is no comparable nexus of programs to deal with restoration of streams, rivers, riparian zones, and floodplains affected by intensification of land use."[118] To fill this gap, the council suggests a goal of restoring 400,000 miles of our degraded riparian ecosystems, about 12 percent of our total river mileage, roughly equal to the number of miles impaired by pollution from point sources and urban runoff. The suggested goal of this effort is to move these ecosystems "as many steps as possible from the negative side of the habitat quality index toward the positive side (through rehabilitation, creation, or full restoration)."[119]

As with the restoration of lakes and wetlands, this effort will not be easy. The NRC panel found that some successful restoration efforts had occurred on smaller streams and on rivers where the headwaters are already protected, but there also have been "many well-meaning, but unsuccessful attempts to restore streams."[120] Restoration of our larger watersheds will be even more difficult. The council notes, consistent with the findings of chapter 2, that river restoration cannot succeed unless we integrate restoration of the river channel and restoration of the adjacent floodplains and riparian habitat, because "rivers and their floodplains are so intimately linked that they should be understood, managed, and restored as integral parts of a single ecosystem."[121]

Conclusion

We return once again to the underlying objective of the Clean Water Act: "to restore and maintain the chemical, physical, and biological integrity of the Nation's waters." All of the concepts discussed in this chapter derive from these few words. Efforts to clean up polluted waters will fail if we do not maintain waters that are still clean. Efforts to rid our waters of chemical pollution will not reap benefits unless we also restore contaminated sediments and lost or degraded habitat. And all of our efforts must be restructured to focus simultaneously on the three parts of the Clean Water Act's objective: chemical, physical, and biological integrity.

Simply repairing or replacing individual programs that are not working as well as planned will not be sufficient to attain this goal. Concerted efforts must be made to integrate these tools into comprehensive programs aimed at restoring the health of whole watersheds. Programs designed to protect individual types of systems, such as lakes, estuaries, and rivers, must be approached from a holistic, integrated perspective. And even these programs must be linked so that efforts to address rivers or wetlands are not at odds with programs to restore their related lakes or estuaries. Only then will we be able to keep our eyes on the target—restoring the health of our nation's waters.

Chapter 8
A National Agenda for Clean Water

All of the problems identified in chapter 4 can be fixed. Some of the necessary changes will be difficult to make, others relatively easy. Some will require more money and personnel, others only more efficient and effective use of existing resources. Many deficiencies can be fixed through administrative action, by changing the regulations, policies, and other instruments of water quality programs. Indeed, the timidity with which the Clean Water Act has been implemented over the past twelve years, along with chronically inadequate resources, explains many of our ongoing failures to meet the most basic goals of the Clean Water Act.

Other issues, however, are more fundamental. Solutions to these basic problems will require changes to the law itself, to broaden, clarify, or redirect the EPA, state, and local activities designed to protect water quality and aquatic ecosystems. Reauthorization of the Clean Water Act, scheduled during the 103d Congress, provides a clear opportunity to make these critical changes to the law.

In this chapter, we outline the major changes needed to get the Clean Water Act back on track—to move us forward quickly toward the zero discharge goal, to give more attention to biological and physical as well as chemical impairment, to protect people and the other species that depend on the health of our aquatic resources, and to achieve our overriding objective to restore and maintain the integrity of the nation's waters.

These recommendations represent a synthesis of the best proposals, developed by a wide range of organizations, on changes needed in the Clean Water Act. Our suggestions draw most heavily on the *National Agenda for Clean Water,* developed by the Clean Water Network, a broadly based network of over four hundred national, regional, and local groups working together to urge Congress to strengthen the Clean Water Act, and on more detailed briefings and issue papers developed by Clean Water Network groups.[1] Other proposals reflect the consensus of an even broader coalition known as Water Quality 2000, a cooperative effort of more than eighty organizations representing industry, all levels of government, academics, professional organizations, and environmental groups to forge a long-term U.S. policy on water quality, reflected in *A National Water Agenda for the 21st Century.*[2] Recommendations drawn from other sources will be noted as appropriate.

We have not organized our recommendations in this chapter to reflect importance—that is, they are not ranked from highest to lowest priority. Rather, we have tracked the order in which we analyzed Clean Water Act programs in chapters 3 through 7. This order proceeds logically from (1) basic tools, such as water quality standards and public information, to (2) whole programs, to address pollution from point sources and polluted runoff, to (3) even broader programs, to eliminate pollution and restore the health of whole aquatic ecosystems, to (4) the need for adequate funding for all of these programs, at the federal, state, and local levels.

We do not repeat our discussion of the serious ecological and programmatic problems discussed in previous chapters. However, as a reminder, and for context, we summarize briefly the problem addressed by each recommendation.

Honing Our Informational Tools

The Need to Expand and Improve Water Quality Standards

Water quality standards are bench marks that define some of the fundamental goals of the Clean Water Act, such as "fishable and swimmable" waters and "physical, chemical and biological integrity." The Environmental Protection Agency is supposed to recommend, and states are supposed to issue, comprehensive standards to define the full beneficial uses of aquatic systems and the water quality necessary to protect these uses. Noncompliance with water quality standards is supposed to drive stricter pollution prevention efforts for point sources, polluted runoff, and other sources of impairment.

This program remains only partially fulfilled. The Environmental Protection Agency has issued only a fraction of the water quality standards it was supposed to complete by 1973, and state implementation is even more sketchy. Where state standards do exist, they often vary widely, so that people, fish, and wildlife are exposed to significantly different levels of toxic and other pollutants. Waters are not protected for fishing and swimming, as the law requires. Furthermore, current EPA and state ambient water quality standards designed to protect "average" consumers do not adequately protect people who consume a lot of fish and shellfish, such as subsistence fishers and Native populations.

The water quality standards program must be revitalized. Congress should convert EPA's recommended water quality standards into enforceable nationwide standards that protect waters for fishing, swimming, drinking, aquatic life and other uses, while preserving the rights of states to enact stricter standards. The agency should be placed on a tight new schedule to complete water quality and sediment standards, including biological water quality standards, for all water bodies and all types of pollutants. In particular, Congress should amend the act to do the following:

1. Expand water quality standards to include biological and wildlife criteria. Water quality standards under the act generally focus on water chemistry and do not adequately ensure the protection, attainment, and maintenance of existing and designated water-body uses. Some states are beginning to implement biological criteria, but these are by no means universal. Section 303 should be amended to clarify that water quality standards are deemed violated whenever activities impair existing or designated uses of water bodies. (Under some present interpretations, standards are considered to be violated only when numeric water chemistry standards are exceeded.) In addition, EPA and the states should be required to issue and enforce numeric wildlife and biological water quality criteria that will protect water-dependent wildlife and whole ecosystems.

2. Ensure that water quality standards are independently enforceable. Point sources violate the act when their discharges exceed permit limits established to maintain water quality standards and are subject to the full range of the act's penalty and enforcement provisions. Sources of runoff pollution and other water-body impairment, however, can cause violations of water quality standards with impunity. Such sources should be subject to enforcement if their actions cause water quality violations, unless they are in compliance with enforceable runoff control and all other applicable requirements of the law.

3. Require EPA to complete criteria for toxic priority pollutants. Congress should expressly require EPA to complete the process of issuing water quality criteria for all of the toxic priority pollutants adopted in 1977. These criteria should be required to address all known or suspected human health or environmental effects.

4. Require EPA to identify and issue criteria for other serious pollutants. Water quality criteria have not been issued for a wide range of pollutants that routinely impair water bodies around the country. Pollutants for which there are no criteria include industrial chemicals, as well as pesticides and nutrients in agricultural runoff; criteria should also be established regarding the sedimentation effects of agricultural runoff. Deadlines should be set for EPA to issue a reasonable number of criteria each year until the task is completed. Dischargers proposing to introduce new chemicals into the nation's waters should be required to submit to EPA prior to discharge all testing information needed for EPA to issue appropriate water quality criteria.

5. Require that EPA issue or revise criteria to address all water bodies. The agency should be required to review all existing criteria and to issue new criteria to address the special needs of water bodies such as lakes, estuaries, and wetlands.

6. Set firm deadlines by which EPA must develop strong sediment quality criteria and standards. Sediment quality criteria and standards must be developed to help protect clean sediments, remediate contaminated sediments, and better manage the disposal of sediments in Great Lake, ocean, and estuarine dump sites. Specific deadlines should be established for EPA to develop strong sediment quality criteria and standards to form the backbone of a national sediment management strategy.

7. Require that EPA water quality criteria apply nationally until states issue their own standards. To avoid serious gaps in state water quality criteria and severe inconsistencies in the degree of protection afforded to citizens and the environment in different states, EPA water quality criteria should apply presumptively in all waters, until superseded by lawful state criteria. State criteria may be more, but not less, strict than those issued by EPA.

8. Require that water quality standards protect the most severely exposed user groups. The practice of setting water quality standards that protect only the "average" person should be prohibited, as this leaves certain groups—such as recreational and subsistence fishers—subject to much higher and inappropriate risks. Standards should be required to protect subpopulations that face higher risk due to higher exposure to pollutants in fish and drinking water.

The Need to Improve Monitoring and Ensure the Public's Right to Know about Routine Water Quality Monitoring

Complete water quality standards are useless without adequate monitoring to determine whether these standards are being met. Yet only a small fraction of waters are monitored routinely either for chemical pollution or for biological impairment. Moreover, monitoring efforts vary from state to state, as do the rules used by the states to evaluate and report which waters meet the requirements of the Clean Water Act and which are safe for fishing, swimming, and other uses. To cure these problems, we recommend that Congress do the following:

1. Direct EPA to issue rules establishing detailed minimum requirements for state water quality monitoring programs. These rules should identify the minimum number and types of pollutants to be measured, including biological and sediment criteria, and protocols for testing frequency and methods. These protocols should ensure that states identify all types of impairment, not just chemical pollution from point sources. Uniform rules should also be established for states to use in deciding and reporting, under section 305(b) of the law and for purposes of public information, which waters fail to meet the goals and requirements of the act.

2. Require states to monitor all waters periodically. States cannot possibly monitor all waters every two years for purposes of state section 305(b) reports. As a result, they often monitor the same rivers, lakes, and estuaries routinely, leaving vast gaps in our knowledge of the condition of other waters. States should be required to divide their waters into groups, ranked from highest to lowest priority, for water quality monitoring. Every two years, one of the groups should receive priority for detailed monitoring and assessment. Other waters can be monitored as necessary to determine the success of control programs. In particular, Congress should add three provisions to the monitoring section of the Clean Water Act, section 305(b):

a. All waters and watersheds in each state must be monitored at a minimum of once every eight years, using the nationally consistent methods, for minimum pollutants and impacts specified by EPA.

b. State water quality monitoring stations and parameters must reflect significant pollution sources, including all major land uses that degrade water quality.

c. States must employ people power—that is, utilize volunteer citizen monitoring teams to help screen waters for runoff-related impacts.

The Need for Public Information

It is often impossible even for experts, much less the general public, to sort through the maze of current monitoring programs. The public has the right to fundamental information about pollution of public waters; answers to the following questions, for example, should be accessible:

● Which waters are safe for fishing, swimming, drinking, and other common uses?

● Who discharges which pollutants into which water bodies, and where?

● Which dischargers are violating their permits?

Even where such information is available, standards for determining which waters should be closed or restricted to protect human health are inconsistent from state to state. The Clean Water Act should be amended to make this basic information available to the public in a usable form. Specifically:

1. All water bodies in which water quality standards are violated or are otherwise unsafe for common uses should be posted at all public access points, in English and in any other prevalent local language. Signs should include information about the health risks posed from swimming, fishing, or other exposure to polluted waters.

2. National standards should be set for closing or restricting waters for fishing and swimming. The Environmental Protection Agency, with the assistance of NOAA for coastal waters, should be required to issue rules establishing methods for determining when to close or restrict waters to fishing or swimming. These rules should establish monitoring frequency and methods as well as criteria for deciding which waters should be closed or restricted and how this information should be communicated to the public.

3. All factories and other facilities that discharge pollutants into public waters should post a sign at the facility entrance informing the public of the discharge and indicating where more information is available. Information on compliance with federal and state permits should be freely available to the public on request and, in the case of public sewage treatment plants, in monthly sewer bills.

4. More and better information should be made available on the use and release of toxic chemicals. The Emergency Planning and Community Right-to-Know Act, enacted as part of the Superfund Amendments of 1986, established the *TRI*, requiring some industries to estimate annual release of certain toxic chemicals to air, land, and water. While highly successful, the right-to-know program has a number of significant limitations. For example, *TRI* today does not cover 40 of the 126 priority pollutants under the Clean Water Act. Additional toxic-releasing industry sectors and "recycling" facilities (such as cement kilns and industrial boilers) and nonmanufacturing industries such as oil and gas wells should be subject to the reporting requirements, as should federal facilities and POTWs. Additional chemicals, regulated under other environmental laws but not currently subject to *TRI* reporting requirements, should be added to the system. Finally, we need improved public access to this vital information source; the current access system is extremely difficult to use.

Closing the Gaps for Point Sources

The 1972 Clean Water Act established a goal that the discharge of toxic and other pollutants be eliminated by 1985. This goal has not been met. Our society continues to use highly toxic chemicals, billions of pounds of which are released into water and other parts of the environment each year, polluting drinking water, contaminating fish, shellfish, and other wildlife, and seriously damaging aquatic ecosystems.

The zero discharge goal has not been met for two reasons. From the start, the EPA and state agencies moved too slowly in implementing and enforcing existing pollution controls. More important, these agencies did not develop a comprehensive program to prevent pollution at its source. Instead, they concentrated solely on treating wastewater before dumping it into rivers, lakes, and coastal waters or shifting it to other parts of the environment.

The next Clean Water Act reauthorization must convert the laudable but vague zero discharge goal into enforceable requirements to eliminate the use of the nation's waters—and other parts of the environment—as dumping grounds for society's wastes. A strategy to reduce and eliminate the use of toxic chemicals in the first place must complement stricter controls on water pollution. The highest priority should be to require industries and other polluters to reduce and eliminate toxics use and generation at the source so that hazards are not simply shifted to workers, consumers, or other parts of the environment, such as air or land. In order to protect aquatic ecosystems and human health, persistent and bioaccumulative pollutants must be identified and phased out of production and use.

The Need to Eliminate the Use and Release of Toxic Pollutants

Human and environmental exposure to toxics occurs not only when these substances are released into air, water, or land through smokestacks, discharge pipes, or lagoons. For example, mercury and lead emissions from solid waste incinerators have been linked to stormwater contamination, as these toxic metals are dispersed into the air and then washed down storm drains. We are also at risk from transportation and on-site accidents and from workplace and consumer exposure to toxic products and building materials.

Solving rather than simply "managing" the problem of human health and water quality impairments from toxics will require a strategy that includes (1) sunsetting (phasing out or banning) the most hazardous substances (those that are most persistent, bioaccumulative, or otherwise toxic) and (2) reducing toxic chemical use in the first instance. In particular, we should:

1. Provide for the sunsetting of particularly harmful toxic contaminants. Certain contaminants are so hazardous to human health or the environment, even in minute quantities, that the only sensible way to address the problems they cause is to remove them from the environment as best we can by phasing them out or banning them. Because there are tens of thousands of chemicals in production, we need a mechanism for singling out the worst of them for this type of regulatory action. The Clean Water Act should require the EPA Administrator expeditiously to establish and apply a procedure for determining which pollutants pose a significant hazard—for example, a procedure that evaluates chemicals on the basis of their persistence or propensity to bioconcentrate or bioaccumulate, their potential harm to reproductive or endocrine processes, or other widespread ecological risks. An amendment should establish a mechanism by which to phase out or ban a chemical identified by this screening procedure within the shortest feasible time. The discharge permitting (or NPDES permitting) program under sections 307(a) and 402 offers the proper vehicle for implementing such phaseouts.

2. Reduce the use and release of toxic contaminants through a new program of pollution prevention planning and reporting by industry. To achieve the zero discharge goal, we need to look beyond the traditional end-of-pipe control approach and consider new ways to use raw products, new manufacturing processes and housekeeping techniques, and new ways to minimize the waste of raw products from the manufacturing process to avoid generating wastes in the first instance. By substituting safe alternatives for toxic chemicals, these new methods should avoid creating new risks to consumers and the environment. Congress should enact a pollution prevention planning and reporting program that can be integrated with the existing *TRI* and CWA reporting programs (such as the 402 permit application and discharge monitoring requirements). Such a provision, which could be adopted as a separate title to the Clean Water Act, would include the following basic elements: (a) annual public reporting on the use of toxic substances within the facility (at the production unit level) and on products (since these may contribute to pollution during postindustrial or consumer treatment, recycling, or disposal); (b) annual public reporting on reduction goals; (c) planning within the facility for achieving pollution prevention goals and for public release of a summary of the plan; (d) technical assistance to industry by EPA; (e) review of federal agency activities for pollution prevention potential; (f) citizen participation; and (g) a fee mechanism to finance program costs.

The Need to Close Loopholes in Existing Pollution Controls

While only a fundamental change of approach accomplished through pollution prevention planning can bring about attainment of the zero discharge goal, fine-tuning our existing system of water pollution controls can bring about more immediate gains. This includes (1) improving the national, technology-based standards written for industrial discharges; (2) closing loopholes that allow discharge of toxic chemicals into the nation's sewage treatment plants without adequate pretreatment; (3) correcting problems that result in weak NPDES permit limits based on both technology and water quality; (4) prohibiting remaining releases of raw or partially treated sewage from CSOs, and pollution from unpermitted sources such as stormwater outfalls; and (5) strengthening citizen enforcement and oversight. Specifically, we propose that Congress do the following:

1. Strengthen the effluent guidelines and categorical pretreatment provisions of the Clean Water Act to increase the focus on pollution prevention and cross-media pollution. Clean Water Act section 304(b)(2)(B) calls on EPA to consider non-water-quality environmental impacts (including energy requirements) in drafting effluent guidelines. However, to date, EPA has been reluctant to establish effluent limitations that specify how to avoid or minimize cross-media environmental impacts. In addition, although section 304(b)(2)(B) gives EPA authority to consider factors such as the "process employed" and "process changes" in setting effluent guidelines, and section 304(b)(3) authorizes EPA to "identify control measures and practices available to eliminate the discharge of pollutants from categories and classes of point sources," to date the agency has been reluctant to use the effluent guidelines process as a mechanism for promoting pollution prevention rather than traditional waste treatment. Congress should bring an end to the toxic shell game by clarifying in the CWA that EPA must select as the BAT that technology which minimizes pollution to all media—and, where feasible, that EPA must prohibit waste-transferring technologies in establishing national guidelines and that it must select technologies, processes, or methods that achieve the best controls via pollution prevention.

2. Expand the scope of the pretreatment program. The Clean Water Act contemplates that industries will pretreat their wastes before discharging them to sewage treatment plants, leaving in their effluent only those wastes that are consistently treated by the POTW. Recent evidence shows that POTWs are not as capable as once thought of consistently treating toxic pollutants. A significant percentage of the toxic pollutants entering sewers comes from commercial and residential sources not covered by the pretreatment program. In many instances, the solution to this problem is to prevent the use of the toxic substance in residential or commercial applications (such as by removing phosphates from laundry detergent).

Three changes are needed to fix this program. First, section 307(b), establishing categorical pretreatment standards, should be clarified to make explicit that indirect dischargers must pretreat all pollutants not actually treated by the POTW to the same degree as that achieved by directly discharging industries. Second, Congress should limit the so-called domestic sewage exclusion in RCRA (which exempts most wastes mixed with domestic sewage from hazardous waste permitting requirements) to pollutants that are covered by national pretreatment standards. Third, a new pollution prevention program should be established, as part of section 307(b), to reduce the release into public sewers of pollutants from commercial and residential products.

3. Close the dilution loophole and other permitting problems in existing law. The 1972 Clean Water Act was supposed to have ended the use of dilution as a solution to pollution. Nearly twenty years later, one of the law's biggest loopholes allows industries and municipalities to continue to dilute their wastes instead of reducing or properly treating them. The use of dilution is based on the mistaken theory that all waters have some capacity to assimilate wastes. Permit limits are calculated assuming that, after wastes mix with some portion of receiving waters, standards will be met at the edge of the mixing zone. While this may be more appropriate for conventional pollutants that degrade, it does not work for toxic chemicals, especially those that persist in the environment or build up in the food chain.

Congress must reinforce the act's original fundamental principles of reducing, preventing, and eliminating pollution by closing dilution and other permitting loopholes. Amendments need to accomplish the following and should apply to all toxic and nonconventional pollutants:

a. Require that all effluent limitations and monitoring requirements be expressed in terms of total mass, in addition to concentration, of each toxic pollutant discharged.

b. Prohibit violations of water quality standards in discharges that are below the level of detection by requiring innovative monitoring techniques.

c. Amend section 301 to expressly prohibit the use of mixing zones (including zones of initial dilution) and dilution with stream flow or lake volume; also require attainment of water quality criteria at the end of the pipe.

d. Amend section 303 to require states to revise their water quality standards and regulations to prohibit the use of mixing zones or dilution. Applicable water quality standards must be met at the end of the pipe. The Environmental Protection Agency should be required to write regulations that will ensure consistency among state rules for writing water-quality–based permits.

e. Where limits are based on plant production technology, require that they reflect actual production, not self-serving company estimates.

The Need to Stop Raw Sewage and Unpermitted Stormwater Discharges

Combined Sewer Overflows

Combined Sewer Overflows cause significant ecological and economic harm. Discharges from CSOs are laden with disease-causing bacteria, leading to the closure of many shellfishing beds and bathing beaches. In addition, the industrial wastes and storm water discharged by CSOs often contain toxic metals and organic compounds that accumulate in sediments and harm aquatic life and the people who catch and eat contaminated fish.

Although the Clean Water Act requires that CSOs be cleaned up, many cities have not yet begun to tackle the job, and EPA has failed to establish a technological standard for the cleanup, despite the fact that the law requires it to do so. Recently, EPA did establish an advisory committee to help develop a consensus resolution to the CSO problem. Parties to these negotiations developed a framework that EPA used in December 1992 to produce draft revised national CSO permit guidance. The guidance calls for both water quality and technical treatment controls for CSOs, elimination of CSOs wherever possible in environmentally sensitive areas, and required levels of treatment.

Even assuming this guidance is finalized, it is not binding regulation. Moreover, it lacks two critical components: (1) firm deadlines by which cities must plan, design, and complete CSO control programs, and (2) additional funding for CSO controls. Still, the committee framework and the resulting EPA permit guidance may form the basis for an unexpected consensus CSO program in the reauthorization of the Clean Water Act. Congress should do the following:

1. **Adopt the essential elements of the consensus CSO framework as part of CWA Reauthorization.**

2. **Add firm but reasonable deadlines by which cities must plan and implement CSO controls.**

3. **Add to state revolving funds for CSO controls.**

Unpermitted Stormwater Discharges

Urban stormwater runoff is a major, uncontrolled source of pollution, carrying oil and grease, nutrients, and heavy metals into our lakes, estuaries, and rivers. Runoff from streets, parking lots, and rooftops rivals both sewage plants and factories as a source of contaminants to lakes, estuaries, and rivers. In 1987, after fifteen years of regulatory delays in implementation of urban and industrial stormwater controls, Congress believed that it had put into place a firm schedule and program for stormwater controls. Section 402(p) of the Clean Water Act, added in 1987, required EPA to issue stormwater permits to all municipalities with populations over one hundred thousand and to all industrial stormwater sources. All stormwater outfalls were to be permitted by October 1, 1992.

Due to additional rule-making delays, however, and loopholes in EPA's regulations, several serious problems remain. First, EPA illegally exempted from permitting several sources of industrial and construction runoff; while a court rejected these loopholes, EPA has yet to issue rules to fill them in. Second, a large number of municipal and other stormwater outfalls remain outside the scope of EPA's program; Congress has set another deadline—October 1, 1994—to fill these gaps. Third, EPA has established no firm stormwater control requirements even for cities covered by the program. To address these problems, Congress should do the following:

1. Amend section 402(p) to specify minimum mandatory practices for all cities to incorporate into their stormwater permits and specify a minimum number of additional practices with a protocol for choosing them from an EPA menu. Additional practices should be chosen based on the following hierarchy: (a) to prevent runoff during new development or redevelopment through water-sensitive site design criteria, including limits on the amount of new pavement and other impervious surfaces and requirements to preserve minimum levels of vegetation; (b) to reclaim and revegetate existing impervious surfaces according to EPA guidelines; (c) to reduce and eliminate chemical contamination of the runoff that does occur via source controls like roofs over gas station pump islands and via coordination with transportation planners to reduce auto emissions; (d) end-of-pipe structural devices like retention ponds and peat-sand filters for parking lot runoff.

2. Give cities stormwater permit provisions that credit them for restoring wetlands that provide natural stormwater functions, such as flood control and filtration.

3. Avoid any further extensions to stormwater permitting requirements and clarify that all remaining sources of urban and industrial storm water require permits by October 1, 1994.

The Need to Strengthen Enforcement and Accountability

Studies by both government and private groups show high levels of permit violations by industrial and publicly owned facilities. Federal, state, and local governments frequently lack the will and the resources to take enforcement action against all significant violators. Government enforcement actions that are brought *are* often ineffective. Compliance schedules are unreasonably long, the penalties imposed do not recover the violator's economic benefit from its noncompliance, and the accuracy of a violator's self-reported monitoring information is not verified to look for unreported violations. Furthermore, EPA's power to impose administrative penalties is capped at $125,000, and courts and EPA cannot impose any penalties on noncomplying federal facilities. The result is a regulatory environment in which it still pays to pollute.

To fill the gaps in government enforcement, hundreds of citizen suits have been filed over the last ten years. A number of problems, however, hamstring citizen enforcement action. Courts have dismissed cases in which past violations have recently ceased, companies have transferred ownership, or states have negotiated prior administrative orders with nominal or no penalties. Citizen groups spend large amounts of time establishing their right to sue, even if they promptly identify members who use waters downstream from illegal discharges. Citizens have difficulty gaining access to permit records to investigate violations, especially for industries discharging into municipal treatment plants. Courts have questioned their own authority to order that penalties be used for environmental mitigation projects. Courts in Virginia have ruled that no one other than the discharger has the right to seek judicial review of discharge permits.

Legal technicalities also impair citizen efforts to ensure that EPA implements its regulatory responsibilities properly. Frequently, when citizens challenge EPA's actions or inactions, the government response is not to defend the case on the merits, but to argue that citizens are in the wrong court. The split and sometimes unclear jurisdictional provisions of the Clean Water Act allow EPA to play these jurisdictional shell games, leaving citizens guessing about where to bring their case.

The nation's clean water program must be made more accountable and more enforceable. Provisions allowing citizens to sue polluters and lax federal and state agencies must be expanded and simplified, and the right of citizens to sue for past violations must be restored. Federal, state, and local oversight roles should be clarified, and portions of the act that are chronically unenforced should be bolstered by new, automatic enforcement provisions, including mandatory minimum penalties.

To make the Clean Water Act more enforceable, and to make government activities more accountable, Congress should do the following:

1. Remove current obstacles to citizen suits. Citizens should be authorized to sue for wholly past violations. The current rule, which prevents citizens from seeking penalties for past violations, makes citizen suits more difficult to bring than EPA or state enforcement actions and encourages polluters to delay compliance until they are notified of a citizen suit. The rule also conflicts with the 1990 amendments to the Clean Air Act, which authorize suits for wholly past violations. Citizens should also be authorized to sue dischargers to POTWs for violations of pretreatment requirements, such as monitoring and reporting requirements, and for violations of local limits set by municipal treatment plants. The standing of citizens to sue should be clarified with a finding that any discharge that violates the act harms citizens who use the receiving waters. And citizens should be allowed to sue for violations of permit limits imposed by court orders.

2. Improve government accountability. Mandatory minimum penalties should be imposed for serious and chronic violations. Significant industrial users of municipal treatment plants should be required to obtain NPDES permits and file monitoring reports with the state and EPA. Public hearings should be required before extended compliance schedules are granted.

3. Strengthen the remedial authority of EPA and the federal courts. As with the 1990 Clean Air Act amendments, judges should have the explicit authority to order that civil penalties be used for projects that enhance public health or the environment. Judges also should have the power to order polluters to take whatever action is needed to remedy the effects of pollution from illegal discharges, including, if appropriate, the cleanup of contaminated sediments. Courts and EPA should be required to assess penalties that, at a minimum, recover the economic benefit that violators obtain by delayed compliance. The existing $125,000 cap on EPA administrative penalties should be removed. Courts and EPA should also have the power to assess civil penalties against federal agencies that violate the act.

4. Increase citizen access to compliance data. Public access to information about toxic chemical releases under the federal right-to-know law has improved public understanding and encouraged pollution prevention efforts. Congress should extend this policy to the Clean Water Act by requiring EPA to make its existing computer database on permit compliance available to the public through computer telecommunication.

5. Increase EPA and citizen oversight of state-issued permits. Congress should clarify that EPA has the authority to issue updated permits and reissue expired permits when states fail to do so in a timely manner. States should be required to allow any person who has participated in the public comment process to obtain judicial review of state-issued permits.

6. Clarify the jurisdiction of federal courts to review EPA actions. The act should be amended to provide that citizens may sue in the U.S. courts of appeals, under section 509, to challenge all final EPA actions that are not subject to citizen suits under section 505. (Section 505 allows citizens to sue to compel the administrator of EPA to perform mandatory duties under the law.) Court of appeals review should include EPA decisions not to issue a particular regulation, in whole or in part, after a request to do so by any party in public comments. To avoid the need to file "protective" appeals, where jurisdiction is still in doubt, the time limit on reviewing EPA actions under section 509 should be tolled when a party files a citizen suit under section 505, until 120 days after a district court rules that it lacks jurisdiction to hear the case.

Preventing Polluted Runoff

More than half of the remaining pollution of the nation's surface waters is caused by polluted runoff from sources such as farms, streets, parking lots, and logging and mining operations. When we use rural and urban land without regard to water quality, we create polluted runoff that destroys aquatic habitat and taints surface water and groundwater with pesticides, manure, lawn fertilizer, sediment, and heavy metals. When we allow these substances to enter our water bodies, we cause fish kills and diseases and smother fish spawning areas, silt up and overenrich our reservoirs, taint our own drinking water, and foul our swimming holes with fecal matter. We can continue to use the land—to farm, graze cattle, harvest timber, and construct buildings—but, without built-in watershed protection practices, destruction of our aquatic resources will continue.

Congress added section 319 to the Clean Water Act in 1987, directing states to (1) assess their waters for runoff damages and (2) create watershed-based programs to repair the damages and prevent further pollution. Unfortunately, implementation of 319 has failed to stem the flow of polluted runoff; the majority of state programs are ineffective and unfocused. The time has come for Congress to prevent polluted runoff through a combination of strict, mandatory programs for activities in polluted watersheds. Specifically, it should do the following:

1. **Reform section 319 of the Clean Water Act to mandate whole-watershed restoration and protection and site-specific pollution control practices in impaired watersheds.** States must create comprehensive target watershed lists that include all significantly degraded and threatened watersheds. These watersheds must be restored on a reasonable timetable.

2. **Require site-level planning and adoption of site-level water quality practices.** As currently required in coastal areas by the Coastal Zone Management Act, all landowners in the target (impaired) watersheds and anyone who breaks new ground for development around the state must bear their fair share of the responsibility for watershed restoration and protection through adoption of runoff measures tailored to each unique site, based on menus of management measures developed by EPA and the states. Each landowner or operator contributing in a target watershed must create a site-level water quality plan consistent with the watershed goals and must implement that plan within a three- to five-year time frame. Technical assistance in writing and enacting the plans, cost-sharing when necessary to help defray farmers' and other landowners' costs, and adequate implementation time should be made available where needed. As water quality monitoring indicates the need, plans

should evolve and practices change over time in an iterative process until the water body is fully restored.

3. Require that federal agencies implement EPA runoff management measure guidance on all lands under their jurisdiction. Significant runoff pollution occurs on lands managed by the BLM, the FS, and other federal agencies. The same type of site-specific management measures and land use practices should be required for activities on federal lands as for those on lands owned by private parties or by state and local governments.

4. Require permits for feedlots and irrigation return flows. These significant causes of water quality degradation in many states are now largely unpermitted and uncontrolled. According to EPA estimates based on the U.S. Census of Agriculture, at least 1.1 million farmers have livestock. Of those, only five thousand to ten thousand operations nationwide are covered by NPDES permits. While the law requires permits for all confined cattle feedlots from which contaminated water is released, EPA established a thousand-unit cutoff for permit issuance. The rest of these pollution sources are largely uncontrolled.

Irrigation, which accounts for 90 percent of the water consumed in the West, results in poisoned return flows that cause serious damage to waters and wetlands, endangering aquatic wildlife with toxics including selenium, boron, molybdenum, and chromium.

Congress should redefine irrigation return flows and clarify the status of smaller confined feedlots and animal waste sources as point sources. The threshold for NPDES permit issuance to feedlots should be lowered to two hundred head of cattle for indirect dischargers and one hundred head for direct dischargers. (Current effluent guidelines use a permitting threshold of one thousand head for indirect and three hundred for direct.)

The current blanket exemption from permitting irrigation return flows should be removed, as recommended by both the NRC and the FWS. While issuing and administering separate permits for each outfall would be inefficient, system-wide permits such as those authorized for storm water under section 402(p) are appropriate and could be used to require system-wide management measures to reduce pollution from irrigation runoff.

Protecting Aquatic Ecosystems and Watersheds

The Need to Keep Clean Waters Clean

The cornerstone of the Clean Water Act is to restore *and maintain* the nation's waters. The states and EPA are charged not only with purifying polluted waters, but with keeping clean waters clean. Certain waters, such as those in national and state parks, wildlife refuges, national forests, wilderness areas, and wild and scenic rivers, are part of our natural legacy and deserve the highest possible protection. Other waters with quality better than required to meet minimum standards must be protected from degradation, except in the most compelling circumstances. This antidegradation principle has always been implicit in the Clean Water Act, and in 1987 Congress added provisions to make it explicit. But these provisions have been misinterpreted or poorly enforced. Pollution of currently clean water continues, and our most pristine waters do not receive the special protection they deserve. National treasures such as the Everglades and Lake Superior, once among our most pristine natural systems, are being lost to pollution and development.

The antidegradation and antibacksliding provisions of the act need to be strengthened. Waters of national and state parks, wildlife refuges, national forests, wilderness areas, wild and scenic rivers, and critical habitat for threatened and endangered species should receive special statutory protection against further pollution. Strict rules for other currently clean waters should be spelled out in the act rather than left to EPA and state discretion, where they often remain ignored or abused. Citizens should have an opportunity to nominate outstanding waters. And Congress should remind EPA that dischargers must be required to move ahead, not backward, in pollution control efforts.

As part of reauthorization of the Clean Water Act, Congress should add the following specific improvements in antidegradation requirements:

1. Require pollution prevention and toxics use reduction before allowing any degradation. Under existing EPA regulations, high-quality waters can be degraded only if a determination is made during the review process that the degradation is "*necessary* to accommodate important economic or social development." Unfortunately, polluters are not uniformly required to show that they have exhausted toxic use reduction and pollution prevention techniques as a condition to establishing necessity. A new Clean Water Act section 303 antidegradation provision should require explicitly that toxics use reduction and pollution prevention principles (TUR/PP) be part of the antidegradation review. A specified hierarchy of relevant pollution prevention measures should be identified, and the party seeking to degrade the water with new or increased sources of pollution should demonstrate that (a) adequate consideration was given to using these measures and (b) implementation of these measures will not avoid or reduce the degradation. If the party cannot meet this burden, degradation should be prohibited. If the burden is met, then the party must implement the TUR/PP measures and satisfy additional antidegradation requirements where the new or increased discharge is not totally averted through use of those measures.

2. Require parties seeking to degrade a water body to demonstrate that new or increased pollutant loadings will not damage ecosystem integrity in the affected waters. Under the present formula, antidegradation review addresses the discharger's alleged necessity to discharge "for social and economic development"—that is, for human uses—but does not assess impairment of ecosystem integrity that may result from the proposed degradation. This amendment would acknowledge the inherent value of ecosystems, independent of human uses, and of the principles of biodiversity and stewardship of the environment.

3. Specify antidegradation review for de minimis discharges of pollutants. The Environmental Protection Agency's current antidegradation policy requires review only of "significant" degradation, which is vaguely defined in terms of demonstrable adverse effects on aquatic ecosystems. As a result, states have widely varying approaches when classifying *de minimis* new or increased pollution, which escapes review even where discharges of toxic, persistent, bioaccumulative chemicals in small amounts would in fact result in serious threats to the environment and human health. Antidegradation review should be required wherever a proposed discharge would increase the water body's load of a toxic, persistent, or bioaccumulative pollutant by any amount. For other pollutants, *de minimis* new or increased discharges could be allowed without triggering the review process so long as the new load would burden the background with a truly minimal amount, defined as no measurable change in ambient water quality. In addition, whenever a decision is made to allow degradation of high-quality waters, a cap should be placed on the cumulative degradation to be allowed in that water body in the future.

4. Apply antidegradation to sources of polluted runoff. Antidegradation policies apply in theory to new or increased sources of polluted runoff. However, few states incorporate antidegradation principles into runoff control programs. Congress should direct EPA to identify all activities that cause runoff pollution and all government programs (as well as statutes and regulations) that authorize, subsidize, or encourage these activities. The CWA and other statutes that regulate these activities should be implemented to require antidegradation plans and procedures. For example, an antidegradation review should be required when a wastewater treatment plant, in order to expand its service area, seeks issuance of a new permit. The proposed increase in surface runoff would have to be justified as necessary, during the antidegradation review, when analyzed in light of pollution prevention principles.

The Environmental Protection Agency's current regulation requires states, before allowing degradation of Tier 2 waters, to "assure that there shall be achieved the highest statutory and regulatory requirements for all new and existing point sources *and all cost-effective and reasonable best-management practices for non-point source control*" (emphasis added). Because the italicized language is so vague, and because polluted runoff programs have been so weak in the past, this requirement has been virtually unenforced. Before degradation of high-quality waters is allowed, the state should be required to develop and begin to implement a watershed-based polluted runoff control plan for the waters in which the proposed or new increased discharge is located, consistent with the previous section, Preventing Polluted Runoff.

5. Require strict protection of pristine waters. One major purpose of the antidegradation program is to preserve Outstanding National Resource Waters (ONRWs)—the last, best, most unique aquatic environments—from harm. Waters in this category should be afforded the highest possible level of protection. Yet few waters receive ONRW protection. A new Clean Water Act antidegradation section should define ONRWs, by presumption, as those waters situated in, or affecting, national or state parks, wildlife refuges, wilderness areas, wild and scenic rivers, estuarine and marine sanctuaries, and critical habitat for endangered or threatened species (unless such designation would be deemed inappropriate by federal land managers). The Environmental Protection Agency should also be required to publish criteria for identification of additional eligible waters based on special ecological, recreational, cultural, or historical significance, after consultation with appropriate federal and state agencies. In addition, a procedure should be established (a) for citizens to petition for designation of water bodies they believe are particularly deserving of ONRW protection and (b) for federal land managers to seek designation of waters in or affecting their areas. In order to provide opportunity for public notice and comment, states should be required to name and formally designate these eligible ONRWs during the Clean Water Act section 303 water quality standards review process or to provide an equivalent opportunity for public review. If any eligible water body is excluded from receiving the ONRW designation, the state doing so should be required to give affirmative justification, with veto and reclassification authority remaining in EPA.

The Need to Strengthen State Water Quality Certification and Other Programs to Protect Biological Integrity

State authority under section 401 of the act to review the water quality impacts of federally regulated activities is not adequate to provide for the restoration and protection of aquatic ecosystems. States currently can waive water quality review of large federally permitted activities, such as major hydroelectric projects. Moreover, because section 401 is limited to federal licenses and permits, and because some courts have restricted section 401 review to point source discharges of chemical pollutants, many activities that seriously impair aquatic ecosystems escape any analysis under the Clean Water Act.

Congress should broaden and strengthen section 401 in the following ways:

1. Apply it to use impairments and other elements of water quality (such as antidegradation) and not just to numeric water chemistry standards.

2. State expressly that it applies (a) to polluted runoff as well as to discharges from point sources, (b) to physical and biological as well as to chemical impacts to wetlands and other water bodies, and (c) to downstream effects.

Congress should also revise or clarify other aspects of the Clean Water Act in order to provide adequate protection of biological integrity.

In particular, Congress should do the following:

1. Incorporate into the NPDES permitting process a new procedure to evaluate effects of proposed discharges on aquatic ecosystems. Whether EPA or the states administer it, this permitting process does not adequately consider the biological effects of authorized point source discharges. The states and EPA should be required to solicit comment from FWS and NMFS (as appropriate) on negative impacts that issuance of proposed NPDES permits will have on physical and biological aquatic habitat in the water body or watershed at issue. As with section 401 certifications, permits should be required to include all appropriate conditions requested by the relevant agency to achieve the goals and objectives of the act.

2. Amend the Clean Water Act to clarify that the Endangered Species Act and other federal environmental laws apply to state actions taken under delegated clean water programs. Under Section 402, states are allowed to assume responsibility for issuing permits for point-source discharges. While the states are exercising a delegated federal responsibility, the question has arisen whether they, in exercising section 402, must abide by federal environmental laws. Section 402 should be amended to clarify that the states must follow federal environmental laws, including the Endangered Species Act, the Fish and Wildlife Coordination Act, and the National Historic Preservation Act.

3. Expressly recognize authority to address effects of low and varying flows on aquatic ecosystems. Dams, diversions, and other water projects have a significant adverse effect on aquatic ecosystem health, since they alter the quantity and timing of flow in rivers, streams, and wetlands. The authority of EPA and the states to address these effects under the act has been questioned. In order to address this problem, section 502(12) of the act, defining discharge of a pollutant, should be amended to add the following: "(C) any release of

water from a point source in amounts, at velocities, or at times, or with such constituents, that prevents the protection, attainment or maintenance of existing or designated uses, or other goals or requirements of the Act." In addition, permits issued under section 402 should be required to prevent low and variable flow impacts through application of water quality standards (expanded to include physical and biological habitat impacts).

The Need to Protect the Physical Integrity of Wetlands and Other Waters

Despite the important role that section 404 has played in reducing the loss and impairment of wetlands and other waters, the law contains limitations and loopholes that allow continued—and unnecessary—destruction of these critical aquatic ecosystems. The only activity regulated by section 404 is the discharge of dredged or fill material; this means that the law does not explicitly cover drainage, excavation, clearing of vegetation, and other wetlands-destroying activities. In addition, agencies that have technical expertise in evaluating wetlands resources, such as FWS and NMFS, do not have an adequate role in commenting on permits issued by the Corps. Furthermore, section 404 allows the Corps to issue "general permits" that give blanket authority to conduct certain activities that can harm wetlands, but there has been insufficient monitoring and oversight to determine if these general permits are leading to substantial cumulative damages to wetlands.

The Clean Water Act—particularly section 404—should be strengthened. Congress should do the following:

1. Establish wetlands protection and restoration as an explicit goal of the Clean Water Act. While implicit in the existing language to restore and maintain all waters, this added step will eliminate any argument that Congress never intended to protect wetlands fully.

2. Ensure that federal procedures for identifying wetlands are based on the best scientific information available. After NAS completes its ongoing study of wetlands delineation, the four federal agencies that implement wetlands programs (EPA, the Corps, the Department of Agriculture, and FWS) should issue a new, improved version of the federal wetlands delineation manual, based on the best available science.

3. Give EPA and the Corps explicit authority to regulate a broader range of wetlands-destroying activities. Section 404 is now limited inappropriately to the discharge of dredge or fill material into wetlands and other waters. This authority should be expanded to include other activities that impair the physical integrity of our waters, such as draining, dredging, excavation, channelization, flooding, clearing of vegetation, driving of pilings or placement of other obstructions, diversion of water, and other actions that change or impair hydrology or physical habitat.

4. Give state and federal resource agencies a more significant role in commenting on permits issued by the Corps. Key resource agencies such as FWS and NMFS now have the same ability as any member of the public to submit nonbinding comments to the Corps before a permit is issued under section 404. Because of the special expertise of these federal agencies, and because of their missions to protect aquatic resources, the Corps should be required to adopt their recommendations or to give written reasons explaining why their recommendations are not necessary to meet the requirements of the act.

5. Require better monitoring and oversight of the general permits authorized under section 404. The vast majority of activities authorized under section 404 do not require individual scrutiny; instead, they fall within the purview of general or nationwide permits. Yet the cumulative impacts of these activities are not evaluated to ensure that long-term harm does not result to important ecosystems and resources. The Corps should be required to review general permits every two years to assess cumulative impacts, and to revoke or revise the permits where necessary to avoid adverse effects.

The Need to Protect Watersheds and Ecosystems

Several recent policy round tables, including Water Quality 2000 and the EPA Office of Water Management Advisory Group, have renewed interest in a watershed-management approach to tackling polluted runoff and other water pollution and aquatic ecosystem problems. To some extent, watershed management is implicit in Clean Water Act programs enacted in 1987 to address particular regional ecosystems, such as the Great Lakes, Chesapeake Bay, Lake Champlain, and National Estuary programs. These efforts should be improved and fully funded. Watershed protection should be a national rather than just a regional enterprise, however, ultimately encompassing every watershed and every water body in the country.

The Need to Strengthen the National Estuary Program

Congress established the National Estuary Program (NEP) under section 320 of the 1987 amendments to the Clean Water Act. The primary purpose of the NEP is to address many of the complex issues—such as habitat protection, polluted runoff, resource management, and land use planning—that have contributed to the significant deterioration of the nation's estuaries. In some respects, the NEP serves as a model for other watershed protection programs around the country.

The NEP authorizes EPA to designate estuaries of national significance. The Environmental Protection Agency then convenes a management conference to address all uses that affect the restoration and maintenance of the chemical, physical, and biological integrity of each estuary. Participants in the management conference include the relevant federal commissions and agencies, the governor and appropriate state commissions and agencies, private companies, organizations, and citizens. The purpose of these five-year conferences is to develop a Comprehensive Conservation and Management Plan (CCMP) to protect and restore the water quality and living resources of estuaries. To date, however, inadequate federal financial commitment to the program has stymied actual implementation of these plans. In order to make NEP a working rather than a paper model for watershed protection, Congress should do the following:

1. Mandate implementation and fixed time frames. Currently, there is no firm requirement that CCMP plans be implemented after they are developed. In addition, the planning process itself is often unnecessarily stalled and extends well beyond its five-year limit. To date, the National Estuary Program has been generally successful at identifying water quality problems. However, the program must move from problem identification to implementation of the solutions to these problems. Section 320 of the CWA should be amended to extend the program, but with strict requirements for the plans to be implemented in a timely way. Federal financial assistance should be provided to assist in the effort. Deadlines are necessary to ensure that individual members of the management conference are not able to stall the entire process.

2. Expand the role of EPA. The role of EPA, as an active participant and as a coordinator of the appropriate environmental agencies, has not been consistent in each of the projects, nor has its level of commitment to the NEP. Section 320 of the CWA should require EPA to take on a more aggressive leadership role in assisting the program to fulfill its goals. Actions taken by EPA need to be coordinated better with the governor's office and state coastal zone management offices. States should also be required to adopt the stronger (or more protective) of their own state coastal management and environmental protection plans or the final CCMP.

3. Increase citizen participation. Citizen participation during the development of the CCMPs is often inadequate, as is the funding necessary to accomplish this goal. Section 320 should be strengthened by requiring citizen participation in all aspects of the CCMP process. Public hearings should be held on a regular basis throughout the life of the program. Funding is needed to ensure full citizen participation and for public education efforts.

4. Establish a Funding Mechanism for State Implementation. Due to state budget shortfalls and a lack of federal support, many states have not been able to follow through on their CCMPs; therefore, there is no guarantee that these plans will ever be implemented, monitored, and enforced. The Clean Water Act should include a funding mechanism to ensure that the states receive federal assistance to implement, enforce, and closely monitor the CCMPs. Federal funds also provide an incentive for states to undertake the more politically difficult task of putting the planning elements into practice. States are eligible to receive CCMP implementation funds under the State Revolving Fund (SRF) program; however, current appropriation levels are severely inadequate to meet the growing demand for funding.

The Need to Expand the Clean Lakes Program

Compared to the NEP, the Clean Lakes Program (CWA section 314) is a token effort, with small amounts of funding for lake assessment and a handful of pilot restoration projects. This program should be either expanded substantially in its own right or woven into a more comprehensive, national program of watershed planning, restoration, and protection (to be discussed later). As for the NEP, any effort to protect lakes on a watershed basis should require (1) comprehensive planning; (2) identification of all sources of impairment and development of solutions to those problems; (3) active participation by citizens and all other involved interests, including EPA and state and local governments; (4) firm deadlines for implementing the identified solutions; and (5) adequate funding.

The Need to Strengthen Ocean Protection Criteria

Despite a specific Clean Water Act provision to protect ocean waters from unnecessary or particularly harmful discharges, massive amounts of pollution continue to be discharged into our coastal waters each year. Unlike lakes and estuaries, it is not possible to protect an entire ocean on a watershed basis. This provision, section 403(c), is designed to assess the full range of impacts on the marine ecosystem, however; in many respects, it is, in theory, a model for whole- ecosystem protection. Unfortunately, EPA's illegal interpretation of section 403(c) allows discharges even where information is lacking to determine whether these comprehensive criteria are met. Congress should reinforce its earlier determination that our marine waters are too important to risk in the face of uncertainty and overrule EPA's illegal rules.

The Need to Protect Significant Regional Ecosystems

Important regional aquatic ecosystems often need special attention from the federal government if they are to gain adequate protection. This is most true of regional resources that cross international, tribal, or state boundaries. This problem has been recognized in the past, and special programs have provided targeted assistance in areas such as the Great Lakes and the Chesapeake Bay. These programs are vital to improving coordination and to addressing water quality problems in these watersheds and must be continued and strengthened.

Similar efforts are needed to accelerate water resource protection and restoration in other significant ecosystems that cross state, international, or tribal boundaries. For example, the Mississippi River has great ecological, economic, and cultural importance to a large number of states but continues to be degraded by sedimentation and other agricultural runoff, toxic industrial pollution, municipal waste, chemical and oil spills, and energy development. Other examples include the Columbia River system and the Rio Grande. Inadequate resources to fund research, coordination, prevention, and cleanup efforts hamper protection of these ecosystems. Because they belong to multiple states or countries, none takes the responsibility to address shared problems. Special programs are needed to direct federal attention and resources to protection of these significant regional ecosystems and to develop appropriate multi-state and international legal authorities.

The Need to Promote Watershed Planning and Implementation on a National Scale

While regional watershed programs are a good start, experts increasingly agree that the time has come to institute (or, in recognition of the section 208 program that was largely abandoned in the 1980s, to reinstitute) watershed-based planning and protection on a national scale, that is, for every major watershed in the United States. The American Planning Association, following on the proposals of Water Quality 2000, is proposing a framework for structuring watershed planning networks, wherein small-scale watershed planning groups are nested inside larger-scale groups around the country. Water Quality 2000 proposed that such nested watershed planning be organized around the existing twenty-one major water resource regions identified by the GS. This framework holds promise for structuring the new state polluted runoff, aquatic ecosystem, and protection programs.

Watershed-based planning can be integrated with many of the other important recommendations in this chapter. Better yet, it could form the principal means of implementing these recommendations. For example, comprehensive watershed planning and management can be used to

1. ensure more comprehensive monitoring for a broader range of impacts on aquatic ecosystems, with the coordinated use of all available monitoring and assessment personnel and resources, including volunteer citizen efforts;
2. coordinate wasteload allocations and water-quality–based permitting of all point sources in each watershed at the same time;
3. develop long-range watershed restoration efforts that address simultaneously all forms of water resource impairment, including the full scope of land uses that generate polluted runoff and other chemical, physical, and biological impairments; and
4. integrate the actions of all levels of government, including federal, state, and local land use planners and managers, into comprehensive watershed protection.

Watershed-based planning and management does not have to entail radical changes to existing programs. Instead, it can be integrated into the existing Clean Water Act system of state water quality programs. For example, as a powerful incentive for watershed approaches, state management grants issued under section 106 of the law, as well as funding for wastewater and other projects under the State Revolving Funds, can be made contingent on the development and implementation of watershed-based plans. Similarly, state monitoring and reporting under section 305(b), as well as the identification and listing of impaired water bodies under sections 304(l), 305(b), and 319, can be integrated into regional systems of watershed-based monitoring, as can state priorities for NPDES permit issuance and other functions under the statute.

The Need to Develop Aquatic Ecosystem Restoration Programs

Unfortunately, protection from future harm is not sufficient to return our aquatic resources to full health. While some effects of pollution dissipate quickly after discharges end, many other impacts—such as contamination of sediments, fish, and wildlife by toxic pollutants—remain for long periods of time. Other impacts, such as widespread physical alterations of aquatic habitat, will remain indefinitely. Given the mandate of the Clean Water Act to restore as well as to maintain the integrity of our aquatic ecosystems, we have both a legal and an ethical mandate to undo as much past harm as is possible. While some Clean Water Act provisions address restoration, absent from the act is a comprehensive program to restore degraded water bodies to full health. The Clean Water Act should be revised to include comprehensive, long-range programs to restore each type of aquatic ecosystem. Congress should do the following:

National Agenda for Clean Water

1. Establish a program of wetlands restoration that retains wetland functions and values. As the contiguous states have lost half of their natural wetlands, and given the severe limits in our knowledge and ability to restore wetlands and other waters, restoration of these critical ecosystems will be difficult. Nevertheless, we must begin, at least, to meet NAS's recommendation to restore 10 million acres of wetlands in the next two decades. Investment in this natural infrastructure will reap long-term benefits in flood control, water quality, fish and wildlife, and other areas.

2. Require EPA to inventory and begin to remediate contaminated sediments. Underwater sediments throughout the country are dangerously contaminated by toxic pollutants dumped by factories, sewage treatment plants, and hazardous waste sites and by mining runoff, polluted agricultural runoff, and other sources. Dredging and dumping sediments, some of which are contaminated, further harm the aquatic environment at more than one hundred dump sites in the ocean and the Great Lakes. A systematic national inventory of contaminated sediment sites has not yet been conducted, but it is widely recognized that each of the four U.S. coasts (Atlantic, Pacific, Great Lakes, and Gulf of Mexico) is experiencing problems from in-place contaminated sediments. Permit applicants for sediment disposal should be required to conduct a waste prevention audit and develop a waste management strategy to protect unpolluted sediments and to stop the continuing pollution of already-compromised sediments.

A comprehensive program is needed to identify and remediate contaminated sediments around the country, beginning with the most polluted hotspots in the Great Lakes and with urban and industrial bays and estuaries. These efforts must be guided by appropriate sediment cleanup standards and methods developed by EPA and given adequate, long-term funding.

3. Create an Urban Watershed Restoration Program as part of the Clean Water Act. Urban streams, rivers, lakes, and bays too often are written off as lost causes. Filled with silt from construction and debris such as cars and tires, and contaminated with high levels of bacteria and toxics from stormwater runoff and raw-sewage overflows, urban waters are often eyesores and public health hazards. Major sections and tributaries of the Anacostia, Chicago, and Los Angeles rivers—and many others—are channelized and banked with concrete or are completely enclosed, transformed into underground sewers. There is a crying need for a national program that empowers urban communities to create their own water restoration programs, marshaling funds, talent, and energy from a variety of sources. This program could restore urban waterways that once were sources of food and recreation, while creating jobs and renewed pride in urban communities.

An Urban Watershed Restoration Program would establish a permanent planning structure that enables local nonprofit or government groups to apply to EPA for basic funding and planning assistance to restore the watershed in their city. These local leaders would provide an outline of their proposed watershed restoration goals, study plan, education, restoration projects, and related work. In turn, EPA would coordinate other federal agencies to assist the local lead group in structuring and funding the work and in training volunteers and other workers in specific project actions such as mapping, forestry, streambank stabilization, reclamation and revegetation of impervious surfaces, and wetlands restoration. Other potential projects could include a sociological survey of urban subsistence fishers.

4. Require restoration as a priority of all watershed planning and protection programs. The Great Lakes, Chesapeake Bay, National Estuary, and Clean Lakes programs, as well as any new watershed-based protection programs created by Clean Water Act reauthorization, should include serious investment in the restoration of the nation's degraded waters and aquatic habitats. Legislative authority for these programs should require the identification of potential restoration projects, and specific funding should be provided to implement them. Congress should adopt as a general goal the restoration targets suggested by NAS—10 million acres of wetlands, 2 million acres of lakes, and 400,000 miles of streams over the next twenty years.

Funding Programs Adequately and Equitably

Historically, Congress has provided significant funding for state and local water pollution control efforts. But unless additional sources of funding are found, our waters will continue to get worse, not better. Congress should provide as much funding as possible for critical treatment programs, such as polluted runoff control and elimination of CSOs. Historically disadvantaged areas may need special attention. But this funding cannot be free. Stricter standards and additional requirements to correct many of the flaws of previous water pollution control efforts must accompany additional federal dollars.

Other sources of funding must be provided, as well. Those responsible for water pollution must pay their full share of pollution control efforts. Direct dischargers should be required to pay the full cost of federal and state permitting and enforcement programs but should not be subject to effluent fees tied to the amount of pollution. Rivers, lakes, coastal areas, and groundwater are public resources and are not for sale at any price. Indirect polluters should be required to pay for polluted runoff control programs through fees on the use of such pollution-generating commodities as fertilizers and pesticides. Cities should fund polluted urban runoff control through such mechanisms as stormwater utilities. At the same time, cities should ensure that low-income citizens have access to essential water and sewer services at affordable rates.

The Clean Water Act is expensive for EPA and the states to administer; furthermore, industry suffers from delay and uncertainty when inadequate resources are available to issue and renew permits promptly. To alleviate these problems, we propose that the CWA be amended to specify that the EPA administrator shall assess a fee to cover the costs of developing effluent guidelines as part of the guideline development process. In addition, a provision calling for a permitting fee program that covers all costs of permit issuance (including compliance monitoring) by EPA or the delegated state should be adopted.

Specifically, Congress should authorize the following funding:

Resources for the EPA Office of Water. The ability of the EPA to conduct necessary research, develop necessary criteria and guidance, and provide necessary oversight and technical assistance is critical to the ability of all levels of government to abate pollution in our nation's waters. The agency must be provided with the resources needed to carry out the federal government's responsibilities under the act. The water quality program has undertaken many new responsibilities since the passage of the 1987 Clean Water Act. Yet due to work force reductions in the early 1980s, it has slightly less staff than it had in 1981. In addition, the program suffered a tremendous cut of 26 percent for FY 1993 in its abatement control and compliance funds used to carry out its water quality responsibilities. (This cut is from funds other than those used for state and local grants and salaries.) Congress should increase resources and staff for the Office of Water.

State and Tribal Water Quality Management Grants. Section 106 of the act authorizes Water Quality Management Grants to the states. These funds assist the states in carrying out their water pollution control responsibilities under the Clean Water Act, such as standards setting, monitoring, technical assistance permitting, and enforcement. The 1987 amendments gave states greater responsibilities for such programs as toxic pollution control, stormwater control, and sludge management. The current authorization is $75 million. The states estimate that in 1992 there was a $400 million shortfall of total funding to meet all their water quality management responsibilities. Congress should authorize an increase of $70 million, for a total of $145 million per year, for section 106 grants. Just as important, Congress should provide this level of funding through the appropriations process.

Congress should also provide adequate funding for tribal water quality programs. State and local governments do not have jurisdiction over water quality on tribal lands; rather, the Clean Water Act was amended to treat tribes as states. But funding has not followed that designation. Currently, EPA dedicates only approximately .1 of 1 percent of its overall Clean Water Act budget to Native Americans nationwide. Congress should ensure that adequate funds are dedicated to tribal water quality programs.

Discharge Permit Fees. Congress should require the establishment of permit fee programs to cover the costs of state and federal permitting, compliance monitoring, and enforcement activities. This is in keeping with the principle that the polluter should pay the cost of cleanup programs. Such a fee system will allow the state and EPA to utilize funds previously used for permitting and enforcement for other water quality management purposes. Any federally mandated permit fee system should have sufficient flexibility so as not to hamper existing state permit fee systems. Fees should be set at a level that reflects the cost to the issuing agency of administering the permit as well as an allocation to a state enforcement program. Permit fees could be set according to the varying complexity of permits, depending on the types and number of pollutants and the number of discharge points. A permit fee system should apply to all NPDES permits.

Polluted Runoff Programs. Although polluted runoff causes more than half of our nation's water quality impairments, it receives a small fraction of all U.S. clean water funds. Congress authorized $400 million for the 319 program for the 1987–92 cycle, but cumulative appropriations through FY 1993 totaled slightly under $200 million, or less than 50 percent of the authorization. A 1990 EPA report projected that annual federal spending on runoff controls would decline to less than 2 percent of the estimated $58 billion in water quality control costs in the year 2000. Clearly, runoff reduction is a neglected stepchild in EPA's budget, in contrast to its role as the number one pollution source.

Funding authorized under section 319 for grants to the states for the implementation of their polluted runoff control programs should be increased to $500 million per year, or $10 million per year per state. Congress should also require stronger state polluted runoff programs in the reauthorization of the Clean Water Act. In addition, it should ensure that EPA has the resources to implement an expanded program adequately. Other sources of funding should also be made available for polluted runoff control. For example, fees on agricultural chemicals could be used to staff technical assistance and outreach positions.

Water Quality Infrastructure. Congress should reauthorize the State Revolving Fund program, increase its funding to a minimum of $5 billion per year, and expand its eligibilities to include CSO abatement and water conservation measures. Funding should not be allowed to pay for infrastructure that encourages sprawl development and degrades water quality. Municipalities should be required to have aggressive water conservation and reuse programs in place, and, except in cases of economic hardship, user fees should reflect the true cost of system financing, operation, maintenance, and replacement. In addition, Clean Water Act reauthorization should address the water quality infrastructure needs of economically disadvantaged communities, providing that they meet a means test for assistance and still pay to the best of their abilities through user fees. Additional revenues to support this increased federal funding could derive from a combination of fees on pollution and activities that cause pollution, so long as fees escalate over time to promote pollution reduction and eventual elimination.

Finally, Congress should provide funds for projects that restore the nation's natural infrastructure—wetlands, forests, rivers, and streams—to help ensure the restoration and protection of aquatic habitat. Such projects can create jobs, enhance environmental values, and increase long-term productivity. In the short term, a large number of jobs can be created quickly by putting people to work in projects such as wetlands reestablishment, stream- and lakeside revegetation, stream channel and bank restoration, and abandoned mine reclamation. These jobs can be targeted to areas of high unemployment, by training out-of-work loggers in the Northwest to restore salmon habitat and inner-city residents in hard-hit urban areas to restore degraded urban waterways (thus saving money by taking these workers off the unemployment roles). Such projects can revive and protect vital economies, such as commercial, recreational, and subsistence fisheries and water-based tourism. By investing now to protect and restore the flood and water quality control benefits of natural wetlands, floodplains, and other waterways, we can save money in the long run by eliminating the need to build more expensive but less effective artificial flood, stormwater, and other pollution controls.

Notes

PART I

1 Congressional Research Service (CRS), *History of the Water Pollution Control Act Amendments of 1972,* ser. 1, 93d Cong., 1st sess. (1972), 137 (hereafter cited as *1972 Legislative History*).

2 The Senate vote was 52 yeas, 12 nays, 36 not voting. CRS, *1972 Legislative History,* 135–36. The House vote was 247 yeas, 23 nays, 160 not voting. Id., 109–112.

3 CRS, *1972 Legislative History,* 164.

4 CRS, *1972 Legislative History,* 117.

CHAPTER 1

1 Theodora E. Colborn et al., *Great Lakes, Great Legacy?* (The Conservation Foundation and the Institute for Research on Public Policy, 1990), xxv.

2 David Zwick and Marcy Benstock, *Water Wasteland* (Bantam Books, 1971), 6, 11, 19–20, 22, 30–33.

3 CEQ, *Second Annual Report* (Washington, D.C.: U.S. GPO, 1971), 107.

4 CEQ, *Second Annual Report,* 219.

5 CEQ, *Second Annual Report,* 218. At that time, the most polluted waters in the country were in the Northeast and Northern plains, and the least polluted were in the Southeast. Id., 220.

6 For a more detailed history of federal water pollution control law, *see* Water Pollution Control Federation (WPCF), The Clean Water Act of 1987, 2d. ed. (1987), 4–9.

7 WPCF, *CWA 1987,* 5–6.

8 CRS, *1972 Legislative History,* 161–62.

9 Clean Water Act (CWA) § 101(a).

10 CWA § 101(a)(1)-(3). Three additional policies included in the 1972 law were actually implementing mechanisms. These included continued and increased federal financial assistance for sewage treatment plant construction, development and implementation of area-wide waste treatment management planning, and increased research and development in pollution control technology. CWA § 101(a)(4)-(6). A seventh policy, to develop and implement programs "for the control of nonpoint sources of pollution," was added in 1987. CWA § 101(a)(7).

11 American Petroleum Institute v. EPA, 540 F.2d 1023, 1028 (10th Cir. 1976).

12 CRS, *1972 Legislative History,* 164.

[13] CRS, *1972 Legislative History,* 1425.

[14] Permits to discharge most pollutants had to be obtained from EPA or a state agency approved by EPA. CWA § 402. Permits to discharge dredge and fill material, including discharges into wetlands, had to be obtained from the U.S. Army Corps of Engineers (the Corps). CWA § 404.

[15] "The Conference agreement specifically bans pollution dilution as an alternative to waste treatment." CRS, *1972 Legislative History,* 166.

[16] Primary treatment involves mechanical screening and settling to remove solids and some organic matter. Secondary treatment uses bacteria in an aerated tank to further break down organic matter. EPA regulations define specific removal levels required for secondary treatment. 40 CFR pt. 133.

[17] CWA § 303.

[18] CRS, *1972 Legislative History,* 1457.

[19] CWA § 502 (19).

[20] PL 95-217, 91 Stat. 1566 (1977).

[21] PL 100-4, 101 Stat. 60 (1987).

[22] Marguerite T. Smith and Debra Wishik Englander, "The Best Places to Live in America," *Money* 20, no. 9 (Sept. 1991), 140.

[23] Roper Organization, *Natural Resource Conservation: Where Environmentalism is Headed in the 1990s* (The Times Mirror Magazines National Environmental Forum Study, June 1992), 5, 8.

[24] Robert E. O'Connor, Richard J. Bord, and Ann Fisher, "Fresh Water Quality, Quantity, and Availability: American Public Perceptions," prepared for the National Geographic Society (Pa. State Univ., 1992), 5, 8.

[25] O'Connor, Bord, and Fisher, "Fresh Water Quality," Fig. I.

[26] Loren Eisely, *The Immense Journey* (Random House, 1946), 15.

CHAPTER 2

[1] EPA, *National Water Quality Inventory, 1990 Report to Congress,* EPA-503/9-92-006 (1992), xxv. While the EPA report identifies 1.2 million U.S. river miles, the National Park Service's *National Rivers Inventory,* places the total almost three times higher, at 3.2 million miles. A. C. Benke, "A Perspective on America's Vanishing Streams," *Journal of the North American Benthological Society* 9 (1990), 77–88.

[2] EPA, *National Water Quality Inventory, 1990,* 134.

[3] World Resources Institute (WRI), *World Resources 1992–1993* (1992), 167.

[4] CEQ, *Twenty-first Annual Report* (1990), 303, 309.

[5] CEQ, *Twenty-first Annual Report*, 309.

[6] CEQ, *Twenty-first Annual Report*, 309.

[7] Ruth Patrick, Emily Ford, and John Quarles, *Groundwater Contamination in the United States*, 2d ed. (Univ. of Pa. Press, 1987), 61–63.

[8] EPA, *National Water Quality Inventory, 1990*, 135–36. These numbers derive from EPA's Needs Surveys for sewage treatment plant construction. Interestingly, EPA's 1980 Needs Survey showed remaining sewage treatment needs of $119 billion–not much higher than the 1990 report. Lynne M. Pollock, "Financing Under the Clean Water Act: The Move from Federal Grants to State Loans," *84 Water Resources Update* (Winter 1991), 25.

[9] NRDC, *When it Rains... It Pollutes* (1992).

[10] Estimates were derived by NRDC based on unit area CSO load factors from: J. Brian Ellis, "Pollutional Aspects of Urban Runoff," in *Urban Runoff Quality*, H.C. Torno, J. Marsalek, and M. Desbordes, eds. (Springer, Verlag, Heidelberg, 1986), 20; and an estimate of total U.S.-CSO service area of 2.5 million acres derived from EPA, "Seminar Publication: Benefit Analysis for Combined Sewer Overflow Control," EPA Office of Technology Transfer (1988).

[11] These data are derived from Bureau of the Census, MA-200(73)-1-(86)-1.

[12] EPA, *Report to Congress: Water Quality Improvement Study*, Office of Water Regulations and Standards (1989), 7–8.

[13] EPA estimates that suspended solids from sewage treatment plants and industry combined have been reduced from almost 10 million tons to about 2 million tons per year. White House statement, (June 1, 1992).

[14] EPA, *1990 Toxics Release Inventory*, public data release, EPA-700-S-92-002 (1990), 15, 55 (hereafter cited as *1990 TRI*).

[15] EPA, *1990 TRI*, 3–5; Deborah A. Sheiman, ed., *The Right to Know More* (NRDC, 1991).

[16] EPA, Decision Document for Martha Prothro, Director, Office of Water Regulations and Standards, on Hazardous Waste Treatment Facilities (undated).

[17] EPA, *Feasibility Report on Environmental Indicators for Surface Water Quality Programs*, EPA Office of Water Regulations and Standards (1990), 105–6.

[18] As explained by monitoring experts in the U.S. Geological Survey (GS):

Despite the expenditure of hundreds of millions of dollars annually on water quality data collection, there is a paucity of data...that are suitable for a scientifically defensible national water quality trends assessment. Much of the data that have been collected are for special purposes, such as intensive surveys, and are not appropriate for documenting changes in water quality over time. Long-term data collection programs have been conducted by many state agencies; however, most of these data are derived from grab samples, which may not be representative of the cross-sectional character of stream quality. In addition, discharge records are not available for many of the state stations, and changes in laboratory procedures used throughout the period of data collection are often not well documented.... However, from the standpoint of longterm trend analysis, the shortcomings of state water quality data are widespread enough to preclude their inclusion in a national data base, which must have reasonably uniform geographic coverage.

D. P. Lettenmaier et al., *Trends in Stream Quality in the Continental United States, 1978–1987* (WRI, 1991), 328.

[19] Lettenmeier et al., *Trends in Stream Quality,* 337.

[20] Lettenmeier et al., *Trends in Stream Quality,* 331.

[21] Richard B. Alexander and Richard A. Smith, "Trends in Lead Concentrations in Major U.S. Rivers and Their Relation to Historical Changes in Gasoline-Lead Consumption," *Water Resources Bulletin* (American Water Resources Association) 24, no. 3 (1988), 568.

[22] Richard A. Smith, Richard B. Alexander, and Kenneth J. Lanfear, *A Graphical Summary of Selected Water-Quality Constituents in Streams of the Conterminous United States, 1980–89* (GS, 1992), open-file report 92-70.

[23] Ruth Patrick, *Surface Water Quality: Have the Laws Been Successful?* (Princeton Univ. Press, 1992), 115–51.

[24] Colborn et al., *Great Lakes, Great Legacy?* xxv.

[25] Colborn et al., *Great Lakes, Great Legacy?* xxvi, 89 (Fig. 4.5); GAO, *Water Pollution: Improved Coordination Needed to Clean Up the Great Lakes,* GAO/RCED-90-197 (1990), 9.

[26] Colborn et al., *Great Lakes, Great Legacy?* xxvi, 89 (Fig. 4.5).

[27] Colborn et al., *Great Lakes, Great Legacy?* xxvi, 91 (Fig. 4.7).

[28] GAO, *Water Pollution: Improved Coordination Needed,* 9.

[29] GAO, *Water Pollution: Improved Coordination Needed,* 9, 32. For a more complete discussion of toxics in the Great Lakes, *see* Colborn et al., *Great Lakes, Great Legacy?* 92–105, 131–46, 165–68.

[30] International Joint Commission (IJC), *Sixth Biennial Report on Great Lakes Water Quality* (1992), 1.

[31] IJC, *Sixth Biennial Report,* 2.

NOTES

[32] Chesapeake Executive Council, *The Chesapeake Bay... A Progress Report 1990–1991* (1991), 5, 12.

[33] Tom Horton and William M. Eichbaum, *Turning the Tide, Saving the Chesapeake Bay* (Chesapeake Bay Foundation, 1991), 100–102.

[34] Horton and Eichbaum, *Turning the Tide*, 76.

[35] William J. Craig and William S. Anderson, "Environment and the River: Maps of the Mississippi," in a report to the McKnight Foundation (corrected version, March 1992), 1, 6–8, 10–11, 24.

[36] Quentin J. Stober and Roy E. Nakatani, "Water Quality and Biota of the Columbia River System," in C. Dale Becker and Duane A. Neitzel, eds., *Water Quality in North American River Systems* (Columbus, Ohio: Batelle Press, 1992), 59–70.

[37] James C. Schmulbach, Larry W. Hesse, and Jane F. Bush, "The Missouri River– Great Plains Thread of Life," in Becker and Neitzel, *Water Quality*, 146–54.

[38] William D. Pearson, "Historical Changes in Water Quality and Fishes of the Ohio River," in Becker and Neitzel, *Water Quality*, 214–21.

[39] 33 USC § 1315(b).

[40] EPA, *National Water Quality Inventory, 1974,* EPA-440/9-74-001; EPA, *National Water Quality Inventory, 1990.* Reports were also issued in 1975, 1976, 1978, 1980, 1982, 1984, 1986, and 1988.

[41] EPA, *National Water Quality Inventory, 1990*, 86–87.

[42] EPA, *National Water Quality Inventory, 1974,* 2. EPA selected these waters based on size (length and volume) and proximity to major population areas.

[43] EPA, *National Water Quality Inventory, 1990*, xxv.

[44] These numbers are derived by multiplying the percentage of waters assessed in each category by the percentage of assessed waters that were monitored. EPA, *National Water Quality Inventory, 1990*, xxv, 4, 18, 48.

[45] EPA acknowledges, however, that "further progress must be made to increase the usefulness of water quality measures reported by the States." EPA, *Guidelines for the Preparation of the 1992 State Water Quality Assessments (305[b] Reports),* Assessment and Watershed Protection Division, Office of Water (1991), 2.

[46] EPA, *Feasibility Report on Environmental Indicators,* 6.

[47] EPA, *Guidelines for the Preparation of the 1992 State Water Quality Assessments,* App. B.

[48] J. R. Karr, "Assessment of Biotic Integrity Using Fish Communities," *Fisheries* 6, no. 6 (1981), 21–27; K. D. Fausch, J. R. Karr, and P. R. Yant, "Regional

Application of an Index of Biotic Integrity Based on Stream-Fish Communities," *Transactions of the American Fisheries Society* 113 (1984), 39–55; J. R. Karr et al., "Assessing Biological Integrity in Running Waters: A Method and Its Rationale" (Champaign, Ill.: Ill. Natural History Survey, 1986), Special Publication no. 5.

[49] EPA, *Feasibility Report on Environmental Indicators*, 82.

[50] Jeffrey Foran, *Regulating Toxic Substances in Surface Waters* (Chelsea, Michigan: Lewis Publishers, 1993), 112–16.

[51] EPA, *Biological Criteria: National Program Guidance for Surface Waters*, EPA-440/5-90-004 (1990).

[52] EPA, *Environmental Indicators for Surface Water Quality Program: Pilot Study*, EPA-905/R-92-001 (1992), 4–17.

[53] EPA, *Feasibility Report on Environmental Indicators*, 83.

[54] NRDC, *Testing the Waters: A National Perspective on Beach Closings* (1992), 4.

[55] Victor Cabelli, *Health Effects Criteria for Marine Recreational Water* (1983), EPA-600/1-84-004, 7, cited in NRDC, *Testing the Waters*, 4.

[56] Constance S. B. Sullivan and Mary E. Barron, "Acute Illnesses among Los Angeles County Lifeguards According to Worksite Exposures," *American Journal of Public Health* 79, no. 11 (Nov. 1989).

[57] EPA, *National Water Quality Inventory, 1990*, 116. Only 31 states provided any information, 18 of which reported 224 beach closure incidents.

[58] From 1988 to 1990, the area surveyed by NRDC was limited to either all or portions of Maine, Mass., R.I., Conn., N.Y., N.J., Del., Md., Fla., and Calif. Only the 1991 total includes all coastal states, with the exception of portions of Mass. In addition, in 1988, several states *did* had not yet begun to maintain records of their beach closures and advisories.

[59] 1989 Calif. data are only for Los Angeles and San Diego counties. 1990 data are for Los Angeles, Mendocino, Monterey, San Diego, San Francisco, San Luis Obispo, Ventura, and San Mateo counties. 1991 data are for all 17 coastal and bay counties.

[60] 1990 Fla. data are only for Dade and Palm Beach counties. 1991 data is for all 35 coastal counties.

[61] The number of closures in Hawaii is an estimate made by the State Dept. of Health, Clean Water Branch, based on county health department closure reports.

[62] 1988–90 Mass. data are only for the area subject to the Metropolitan District

NOTES

Commission's jurisdiction. 1991 data are from the MDC, the Lynn-Swampscott area, Quincy, Hull, Plymouth, and Crane's Beach.

[63] Monitoring in Miss. showed that 9 of 11 beach sites exceeded state bacteria standards for summer geometric averages. No closings or advisories were issued at this time.

[64] 1990 and 1991 totals for NYC include 48-hour advisories against swimming after every rainfall event in excess of .4 inches per day. The NYC Department of Health issued an annual rainfall advisory by press release for Locust Point, Little Neck Bay, Coney Island, and Seagate in 1990, and for Coney Island and Seagate in 1991. Based on this advisory, NRDC estimated 72 rainfall advisories for 1990 and 54 rainfall advisories for 1991, using NOAA climate data for Central Park.

[65] NRDC, *Testing the Waters*, 3; EPA, *National Water Quality Inventory, 1990*, 54, 58.

[66] Zwick and Benstock, *Water Wasteland*, 31, citing *New York Times* (July 9, 1970).

[67] John F. Dwyer, *Outdoor Recreation Participation: Blacks, Whites, Hispanics, and Asians in Illinois* (U.S. Dept. of Agriculture, Forest Service [FS], 1991). *Also* John F. Dwyer and Ray Hutchison, "Outdoor Recreation Participation and Preferences for Black and White Chicago Households," in Joanne Vining, ed., *Social Science and Natural Resource Recreation Management* (Westview Press, Boulder, CO).

[68] Dwyer, *Outdoor Recreation Participation*, 2.

[69] Christine Ruf and Jamal Kadri, *Combined Sewer System Communities: Income and Ethnicity Inequities*, prepared for the Office of Policy, Planning and Evaluation, Water Policy Branch (EPA, 1992), 4–5.

[70] David K. Gordon and Robert F. Kennedy, Jr., *The Legend of City Water: Recommendations for Rescuing the New York City Water Supply* (The Hudson Riverkeeper Fund, Sept. 1991), 30.

[71] *The New York Times* (April 20, 1993), C3.

[72] Gordon and Kennedy, *Legend of City Water*, 31.

[73] M.W. LeChevallier, W.D. Norton and R.G. Lee, "Occurrence of *Giardia* and *Cryptosporidium* spp. in Surface Water Supplies," *Applied and Environmental Microbiology* 5, no. 9, 2610-16 (1991).

[74] W. C. Levine, W. T. Stephenson and G. F. Craun, "Waterborne Disease Outbreaks, 1986–1988," *Morbidity and Mortality Weekly Report* 39, no. SS-1 (1990), CDC Surveillance Summaries; B. L. Herwaldt, G. F. Craun, S. L. Stokes and D. D. Juranek, "Outbreaks of Waterborne Disease in the United States: 1989–90," *Journal of the American Waterworks Association*, April 1992, 129–35.

NOTES

75 G. F. Craun, "Surface Water Supplies and Health," *Journal of the American Water Works Association* 40, no. 49 (Feb. 1988), *cited in* N. L. Dean, *Danger on Tap: The Government's Failure to Enforce the Federal Safe Drinking Water Act* (National Wildlife Federation, 1988), 15.

76 52 Fed. Reg. 42181, 42183 (1987).

77 For purposes of evaluating the success of the Clean Water Act, which addresses surface water but not groundwater quality, public drinking water systems whose supply is taken from groundwater must be distinguished from those taken from surface water. Considerable evidence points to pollutants in groundwater from wells around the country. For example, EPA's *National Survey of Pesticides in Drinking Water Wells* (1992), xiv–xv, estimated that approximately 19 million people are exposed to nitrates in rural drinking water wells, with about 1.5 million people (and 22,500 infants under one year old) exposed to levels above the legal limit. (Infants are particularly susceptible to nitrates in drinking water, which may cause blue baby disease.) Moreover, more than 10 percent of all community water supply wells and more than 4 percent of all rural domestic wells contain detectible levels of one or more pesticides (although a much smaller number of wells are estimated to have pesticides above levels set to protect human health). These figures indicate that groundwater contamination should be prevented through national regulation, either as part of the Clean Water Act or through another statute or program.

78 Elizabeth Reichheld, Lowell Ungar, and David Loveland, eds., *Crosscurrents: The Water We Drink–A Report on a Survey of Drinking Water Utility and State Officials* (League of Women Voters Education Fund, 1989).

79 GAO, *Drinking Water: Compliance Problems Undermine EPA Program as New Challenges Emerge*, a report to the Chairman, Environment, Energy, and Natural Resources Subcommittee, Committee on Government Operations, House of Representatives, GAO/RCED-90-127 (1990), 63–65.

80 33 USC § 300j-4.

81 EPA, *The National Public Water System Supervision Program–FY 1991 Compliance Report* (1992), 16–22.

82 Erik Olson, Senior Attorney, NRDC, personal communication (May 10, 1993).

83 N.L. Dean, *Danger on Tap, FY 1988 Update* (National Wildlife Federation, 1989), 5.

84 "Is Our Fish Safe to Eat?" *Consumer Reports* 57, no. 2 (Feb. 1992), 103.

85 "Is Our Fish Safe to Eat?" 112.

[86] EPA, *Assessing Human Health Risks From Chemically Contaminated Fish and Shellfish: A Guidance Manual* EPA-503/8-89-002 (1989), 1.

[87] F. E. Ahmed, ed., *Seafood Safety,* Institute of Medicine, Committee on the Evaluation of the Safety of Fishery Products, Food and Nutrition Board (Washington, D.C.: National Academy Press, 1991).

[88] For a more detailed discussion of this phenomenon, *see* Foran, *Regulating Toxic Substances,* 56–58.

[89] EPA, *National Water Quality Inventory, 1990,* 90–92.

[90] EPA, *National Water Quality Inventory, 1990,* 91. Chapter 4 addresses in more detail the variation in state criteria for the fish advisories.

[91] EDF, *The Contamination of Our National Waters: A Report on State Fish Advisories and Bans in the United States* (1992).

[92] EPA, *Assessing Human Health Risks,* 4; Ahmed, *Seafood Safety,* 240–51; Foran, *Regulating Toxic Substances,* 62–67.

[93] EPA, *National Study of Chemical Residues in Fish,* EPA-823/R-92-008(a) (1992). The principal findings are included in the Executive Summary of a 2-volume report. Id., xv–xxiv.

[94] C. J. Schmitt and W. G. Brumbaugh, *National Contaminant Biomonitoring Program: Concentrations of Arsenic, Cadmium, Copper, Lead, Mercury, Selenium, and Zinc in the U.S. Freshwater Fish, 1976–1984* (1990). *Also* C. J. Schmitt, J. L. Zajicek, and P. H. Peterman, *Residues of Organochlorine Chemicals in U.S. Freshwater Fish, 1976–1984* (1990). Prepared for the National Contaminant Biomonitoring Program, National Fisheries Contaminant Research Center, FWS, U.S. Dept. of Interior. *Archives of Environmental Contamination and Toxicology* 19 (1990).

[95] EPA, *Assessing Human Health Risks,* Tables H-1 and H-2.

[96] Thomas P. O'Connor, ed., *Coastal Environmental Quality in the United States, 1990: Chemical Contamination in Sediment and Tissues* (NOAA, 1990), 11.

[97] O'Connor, *Coastal Environmental Quality,* 11; Thomas P. O'Connor, T.P., "Concentrations of Organic Contaminants in Mollusks and Sediments at NOAA National Status and Trend Sites in the Coastal and Estuarine U.S," *Environmental Health Perspectives* 90 (1991), 70, 73. National Status and Trends Program, Ocean Assessments Division, NOAA.

[98] Summaries of the Status and Trends and Mussel Watch Programs were derived from Thomas P. O'Connor, "Concentrations of Organic Contaminants in Mollusks and Sediments"; O'Connor, *Coastal Environmental Quality;* NOAA, *A*

Summary of Data on Tissue Contamination from the First Three Years (1986–1988) of the Mussel Watch Program–Progress Report, NOAA Technical Memorandum NOS OMA 49 (1989); G. G. Lauenstein, A. Robertson, and T. P. O'Connor, "Comparison of Trace Metal Data in Mussels and Oysters from a Mussel Watch Programmed of the 1970s with Those from a 1980s Programme," *Marine Pollution Bulletin* 9 (1990).

99 NOAA, *The 1990 National Shellfish Register of Classified Estuarine Waters* (1991), 5. The intermediate categories of waters in which shellfishing is restricted or conditional remained about equal.

100 Ahmed et al., *Seafood Safety,* 31.

101 Ahmed et al., *Seafood Safety,* 31, 35.

102 For example, NOAA's Mussel Watch Program found a strong statistical correlation between levels of contaminants in mussels and the associated sediment. O'Connor, "Concentrations of Organic Contaminants," (1991), 70.

103 Beth Millemann and Eleanor Kinney, eds., *Getting to the Bottom of It: Threats to Human Health and the Environment from Contaminated Underwater Sediments* (Coast Alliance, 1992).

104 Millemann and Kinney, *Getting to the Bottom of It,* 1.

105 Arthur D. Little, *An Overview of Sediment Quality in the U.S.* (1987).

106 Little, *Overview of Sediment Quality,* 19–22, 42–45, 47, 57–60.

107 NRC, *Contaminated Marine Sediments–Assessment and Remediation* (1989), 1.

108 NRC, *Contaminated Marine Sediments,* 4.

109 EPA, *Environmental Equity: Reducing Risk for All Communities,* EPA-230/R-92-008A (1992), 12.

110 Patrick C. West, "Invitation to Poison? Detroit Minorities and Toxic Fish Consumption from the Detroit River," in Bunyan Bryant and Paul Mohai, eds., *Proceedings of the Michigan Conference on Race and the Incidence of Environmental Hazards* (1992).

111 West, "Invitation to Poison?" 124.

112 West, "Invitation to Poison?" 124.

113 EPA, *Environmental Equity: Reducing Risk,* 12. The survey shows that Asians eat 21 g/person/day, while African Americans eat 16 g/person/day and Caucasians eat 14.2 g/person/day. EPA, *Exposure Factors Handbook* (1990), 2–34.

114 J. R. Karr, "Assessment of Biotic Integrity Using Fish Communities," *Fisheries* 6, no. 6 (1981): 21–27.

NOTES

[115] Larry Master, "The Imperiled Status of North American Aquatic Animals," *Biodiversity Network News* 3, no. 3 (1990), 7. The Nature Conservancy.

[116] Master, "Status of North American Aquatic Animals," 7. The figures are mammals (6%), birds (5%), reptiles (6%), amphibians (3%), fishes (7%), crayfishes (1%), and mussels (11%).

[117] T. R. McClanahan, "Viewpoint: Are Conservationists Fish Bigots?" *Bioscience* 40 (1990), 2.

[118] R. Dana Ono, James D. Williams, and Anne Wagner, *Vanishing Fishes of North America* (Stone Wall Press, 1983), 7.

[119] David E. Blockstein, "An Aquatic Perspective on U.S. Biodiversity Policy" (draft accepted for publication in *Fisheries*, 1992), 6.

[120] Ono et al., *Vanishing Fishes,* 231.

[121] Ono et al., *Vanishing Fishes,* 231; Deacon et al., "Fishes of North America Endangered, Threatened, or of Special Concern," *Fisheries* (March-April 1979), 29–44.

[122] Ono et al., *Vanishing Fishes,* 20–24, 37–38, 88–91, 175–88, 210.

[123] Ono et al., *Vanishing Fishes,* 10, 33, 37–38, 88–91, 109, 175–76.

[124] Jack E. Williams et al., "Fishes of North America Endangered, Threatened, or of Special Concern: 1989," *Fisheries* 14, no. 6 (1989), 12. 123 of the species are in Mexico, and 23 in Canada.

[125] Williams et al., "Fishes of North America," 12.

[126] Robert R. Miller, James D. Williams, and Jack E. Williams, "Extinctions of North American Fishes During the Past Century," *Fisheries* 14, no. 6 (1989), 22, 34.

[127] The House Report defined *integrity* as "a condition in which the natural structure and function of ecosystems is maintained" and *natural* as "levels believed to have existed before irreversible perturbations caused by man's activities." CRS, *1972 Legislative History,* 76–77. The Senate report was even more explicit, instructing that *integrity* be determined by reference to "historical records on species composition." Id., 1468.

[128] 50 CFR § 17.11 (July 15, 1991, supplemented as of March 15, 1992).

[129] 56 Fed. Reg. 58804 et seq. (Nov. 21, 1991).

[130] Karr, Toth, and Dudley, "Fish Communities of Midwestern Rivers: A History of Degradation," *BioScience* 35, no. 2 (1985). One of the authors believes this is typical of major U.S. river systems. Dr. James Karr, personal communication (Dec. 8, 1992).

[131] Benke, "America's Vanishing Streams," 77–88.

[132] Walter V. Reid and Mark C. Trexler, *Drowning the National Heritage: Climate Change and U.S. Coastal Biodiversity* (World Resources Institute, undated), 14–16.

[133] "U.S. Coastal Habitat Degradation and Fishery Declines," in J. R. Chambers, ed., *Opening Sessions of the Fifty-seventh North American Wildlife and Natural Resources Conference* (NOAA and NMFS, 1991), 3–4.

[134] Karr, Toth, and Dudley, "Fish Communities of Midwestern Rivers," 93.

[135] CEQ, *Environmental Trends* (1989), 106.

[136] FWS, U.S. Dept. of Interior, *1992 Status of Waterfowl and Fall Flight Forecast*, 22–23.

[137] Data supplied by Bruce Peterjohn, Coordinator, Breeding Bird Survey.

[138] EPA, *Fish Kills Caused by Pollution: 1977–1987*, Office of Water Regulations and Standards (1987). *Also* EPA, *Fish Kills Caused by Pollution: Fifteen-Year Summary, 1961–1975*, EPA-440/4-78-001 (1978).

[139] EPA, *National Water Quality Inventory, 1990 Report*, 114.

[140] EPA, *Fish Kills Caused by Pollution: 1977–1987*, v.

[141] This estimate reflects adjusted numbers of fish killed as reported to EPA, normalized to reflect states that did not report; and a doubling of the fish killed as reported to NOAA, to reflect the fact that only data from the 1980s are available. The estimates are likely to be highly conservative because EPA assumes that many fish kills go unreported, and counts are conservative for kills that are reported; and because the NOAA data show a decrease over time in the number of fish killed, meaning that more fish were likely to have been killed in the 1970s than in the 1980s. Overlap between EPA and NOAA data appears minimal, as only a small fraction of spills reported to EPA were in coastal waters (for example, less than 300 out of almost 9,000 incidents [3%] between 1961 and 1975). This is less than or equal to the total number of coastal fish kills reported by NOAA in a single year between 1980 and 1989.

[142] Mehrle et al., "Toxicity and Bioconcentration of 2,3,7,8-Tetrachlorodibenzo-dioxin and 2,3,7,8-Tetrachlorodibenzofuran in Rainbow Trout," *Environmental Toxicology and Chemistry* 7 (1988), 47–62.

[143] NWF and the Canadian Institute for Environmental Law and Policy (CIELAP), *A Prescription for Healthy Great Lakes: Report of the Program for Zero Discharge* (1991), 4–5. More detailed information is available from the results of a 1991 symposium on the effects of toxics in the Great Lakes. Michigan Audubon Society, *Cause-Effect Linkages II Symposium Abstracts*, prepared for a symposium sponsored by the Michigan Audubon Society (Sept. 27–28, 1991).

[144] NWF and CIELAP, *A Prescription for Healthy Great Lakes,* 5.

[145] NOAA, "Coastal Habitat Degradation," 5–6.

[146] NOAA, "Coastal Habitat Degradation," 5–6.

[147] NOAA, "Coastal Habitat Degradation," 6.

[148] J. David Allan and Alexander S. Flecker, "Biodiversity Conservation in Running Waters" (unpublished) (1991), 9.

[149] Master, "Status of North American Aquatic Animals," 7; Ono et al., *Vanishing Fishes,* 5; Allan and Flecker, "Biodiversity Conservation," 24.

[150] R. D. Judy, Jr., et al., *1982 National Fisheries Survey, Volume I Technical Report: Initial Findings,* FWS/OBS-84/06 (FWS, U.S. Dept. of Interior, 1984), viii.

[151] Judy et al., *1982 National Fisheries Survey,* 22 (Table 8), 25 (Table 10), 28 (Table 11), 46 (Table 18).

[152] Judy et al., *1982 National Fisheries Survey,* 46–47. Predictions over the next five years were that 41,000 miles would degrade from rank 3 to 2; the percentage of streams that would not support any fish would increase from 23.1 to 24 for all streams and 3.1 to 3.5 for perennial streams; percentage of streams with a rank of 1 would increase from 9.7 to 12 (all) and from 5.2 - 7.4 (perennial); and percentage of streams with a rank of 3 would decline from 25.1 to 22.2 (all) and 23.9 to 21.1 (perennial). Since this important work was terminated, the accuracy of these predictions cannot be tested.

[153] EPA, *National Water Quality Inventory, 1990 Report,* 14.

[154] NWF, *Endangered Species, Endangered Wildlife: Life on the Edge* (1992), 3.

[155] NWF, *Endangered Species,* 7–8.

[156] T. E. Dahl and C. E. Johnson, *Status and Trends of Wetlands in the Conterminous United States, Mid-1970's to Mid-1980's* (Washington, D.C.: FWS, U.S. Dept. of Interior, 1991).

[157] Dahl et al., *Status and Trends of Wetlands,* 3.

[158] Dahl et al., *Status and Trends of Wetlands,* 8. Of the remaining wetlands, 97.8 million acres (95%) are freshwater (inland) and 5.5 million (5%) are estuarine (coastal) wetlands.

[159] Dahl et al., *Status and Trends of Wetlands,* 10–11.

[160] Dahl et al., *Status and Trends of Wetlands,* 12–14.

[161] EPA, *National Water Quality Inventory, 1990,* 66.

[162] One source estimates that 80% of the nation's remaining wetlands are in floodplains. John H. McShane, "Integrating Provisions of the National Flood Insurance Program with Multi-Objective River Corridor Management" (1992),

in manuscript to be published in the *Proceedings of the Sixteenth Annual Conference of the Association of State Floodplain Managers,* 2.

[163] Allan and Flecker, "Biodiversity Conservation," 12.

[164] Federal Interagency Floodplain Management Task Force, *Floodplain Management in the United States: An Assessment Report,* vol. 1, Summary, 11–12. Prepared by the Natural Hazards Research and Applications Information Center, Univ. of Colo. at Boulder (1992), contract no. TV-72105A.

[165] F. L. Knopf et al., "Conservation of Riparian Ecosystems in the United States," *Wilson Bulletin* 100, no. 2. (June 1988), 273–74.

[166] Floodplain Management Task Force, *Floodplain Management,* 11–12.

[167] Floodplain Management Task Force, *Floodplain Management,* 16, 18–19, 21.

[168] B. L. Swift, "Status of Riparian Ecosystems in the United States," *Water Resources Bulletin,* paper no. 84007, vol. 20, no. 2 (April 1984), 224.

[169] Swift, "Status of Riparian Ecosystems," Tables 1–3, 224–25.

[170] CEQ, *Environmental Trends,* 27–28.

[171] Swift, "Status of Riparian Ecosystems," 224.

[172] NOAA, "Coastal Habitat Degradation," 21.

[173] Benke, "America's Vanishing Streams," 77–88.

[174] Benke, "America's Vanishing Streams," 78.

[175] Benke, "America's Vanishing Streams," 84. The greatest number of high-quality rivers was found in the Southeast, but this region also had the smallest percentage of rivers in a protected status. Id., 83.

[176] Benke, "America's Vanishing Streams," 83.

[177] Benke, "America's Vanishing Streams," 87.

[178] Benke, "America's Vanishing Streams," 80 (fig. 2).

[179] Benke, "America's Vanishing Streams," 81.

[180] Benke, "America's Vanishing Streams," 81.

[181] NOAA, "Coastal Habitat Degradation," 2.

[182] NOAA, "Coastal Habitat Degradation," 1.

[183] Thomas J. Cuilliton et al., eds., *The Second Report of a Coastal Trends Series, 50 Years of Population Change along the Nation's Coasts, 1960–2010* (NOAA, 1990), 41 pp.

[184] NOAA, "Coastal Habitat Degradation," 3–5.

[185] NOAA, *Building Along America's Coasts. 20 Years of Building Permits, 1970–1989* (1992), 8–9.

CHAPTER 3

[1] NOAA, *Fisheries of the U.S., 1990,* Current Fishery Statistics No. 9000 (1991), 7.

[2] R. N. Sampson and D. Hair, *Natural Resources for the Twenty-first Century,* American Forestry Association (Island Press, 1990), 236.

[3] NOAA, *Our Living Oceans, The First Annual Report on the Status of U.S. Living Marine Resources* (1991), 8.

[4] NOAA, *Fisheries of the U.S., 1990,* 7.

[5] NOAA, *Fisheries of the U.S., 1989,* Current Fishery Statistics No. 8900 (1990), 26.

[6] FWS, U.S. Dept. of Interior, *1985 National Survey of Fishing, Hunting, and Wildlife Associated Recreation* (1988), 14.

[7] EPA, *Clean Water and the American Economy* (1992), 6-2, n.3.

[8] EPA, *Clean Water and the American Economy,* 14.

[9] FWS, *1985 National Survey,* 15.

[10] NOAA, 1991, *Recreational Shellfishing in the United States, Addendum to the 1985 National Survey of Fishing, Hunting and Wildlife-Associated Recreation* (1991), 2, 11.

[11] FWS, *1985 National Survey,* 150.

[12] U.S. Dept. of Commerce, *Statistical Abstracts of the United States,* 11th ed. (1991), 230, 235, 239.

[13] Sampson and Hair, *Natural Resources,* 205, 207.

[14] Sampson and Hair, *Natural Resources,* 261.

[15] FWS, *1985 National Survey,* 44.

[16] EPA, *Clean Water and the American Economy,* 6–14, citing Frederick Bell and Vernon Leeworthy, "Recreational Demand by Tourists for Saltwater Beach Days," *Journal of Environment, Economics and Management* 18 (1990), 189–205.

[17] Dept. of Commerce, *Statistical Abstracts,* 228, 237.

[18] EPA, *Clean Water and the American Economy,* 6-9, 6-12 (exhibit 6–7), citing National Marine Manufacturers Association, "The Importance of the Recreational Marine Industry" (1990).

[19] EPA, *Clean Water and the American Economy,* 6-14, citing Frederick Bell and Vernon Leeworthy, *An Economic Analysis of the Importance of Saltwater Beaches in Florida* (Florida Sea Grant College, Feb. 1986).

[20] Mark Sagoff, *The Economy of the Earth: Philosophy, Law, and the Environment* (Cambridge Univ. Press, 1988), chap. 4.

21 Richard T. Carson and Robert Cameron Mitchell, *The Value of Clean Water: The Public's Willingness to Pay for Boatable, Fishable, and Swimmable Water* (unpublished) (1991).

22 Carson and Mitchell, *Value of Clean Water,* 26–28.

23 EPA, *Clean Water and the American Economy,* 6–17.

24 F. W. Bell, Application of Wetland Valuation Theory to Florida Fisheries (Florida Sea Grant College, 1989), cited in NOAA, *Estuaries of the United States, Vital Statistics of a National Resource Base* (1990), 6–7.

25 EPA, *America's Wetlands: Our Vital Link,* Office of Wetlands Protection, Office of Water, OPA-87-016 (1988), 5.

26 Frances R. Thibodeau and Bart D. Ostro, "An Economic Analysis of Wetland Protection," *Journal of Environmental Management* 12 (London: Academic Press, 1981), 21–27.

27 Md. Dept. of Economic and Employment Development, Office of Research, *Economic Importance of the Chesapeake Bay* (1989), 1–4.

28 The Sounds Conservancy, *Perspective on Shell Fisheries in Southern New England* (1992), 7–8. The breakdown by state is as follows: Mass., $17 million; R.I., $2.5–$5.5 million; Conn., $12 million; N.Y., $2–$4 million.

29 EPA, *Clean Water and the American Economy,* 5–11 to 5–14.

30 Floodplain Management Task Force, *Floodplain Management,* 8.

31 EDF and World Wildlife Fund (WWF), *How Wet Is a Wetland? The Impacts of the Proposed Revisions to the Federal Wetlands Delineation Manual* (1992), 36.

32 EDF and WWF, *How Wet is a Wetland?* 37.

33 Floodplain Management Task Force, *Floodplain Management,* 18–19.

34 NRC, *Managing Coastal Erosion,* Committee on Coastal Erosion Zone Management, Water Science and Technology Board (National Academy Press, 1990), 37, 79.

35 EPA, *Pesticides and Ground-Water Strategy: A Survey of Potential Impacts.* Office of Pesticide Programs, Biological and Economic Analysis Division (1991), 5–6.

36 Gordon and Kennedy, *Legend of City Water,* 30.

37 GAO, *Drinking Water: Compliance Problems Undermines EPA Program as New Challenges Emerge,* GAO/RCED-90-127 (1990), 53.

38 EPA, *Clean Water and the American Economy,* 6–8.

39 Douglas D. Ofiara and Bernard Brown, "Marine Pollution Events of 1988 and Their Effect on Travel, Tourism, and Recreational Activities in New Jersey,"

Bureau of Economic Research, Rutgers Univ., and Marine Fisheries Administration, Fish Game, and Wildlife, N.J. Dept. of Environmental Protection (March 1989).

[40] van der Leeden et al., *The Water Encyclopedia*, 2d ed. (Chelsea, Mich.: Lewis Publishers, 1990), 541.

[41] William H. Bruvold, Betty H. Olson, and Martin Rigby, "Public Policy for the Use of Reclaimed Water," *Environmental Management* 5, no. 2 (1981), 95.

[42] According to one estimate, it would take 53 million barrels of oil to replace with petroleum-based fertilizers the amount of nutrients disposed of annually in U.S. wastewater. Sandra Postel, "Conserving Water: The Untapped Alternative," *Worldwatch* 67 (1985), 35.

[43] Marcia Lowe, "Down the Tubes," *Worldwatch* (March–April 1989).

[44] Bruvold, Olson, and Rigby, "Public Policy," 98.

[45] Bruvold, Olson, and Rigby, "Public Policy," 47.

[46] For example, the cost of treating water to the level appropriate for park irrigation is currently approximately $.55 per 3,785 liters [1981] (1,000 gallons). The cost of disinfected secondary effluent, the federal minimum, is currently $.32 per 3,785 liters. The marginal cost of water treated to a level appropriate for park irrigation is therefore $.23 per 3,785 liters. Orchard irrigation water, for which secondary treatment is sufficient, has a marginal treatment cost of $.00 per 3,785 liters because $.32 minus $.32 equals $.00 when secondary treatment is mandatory. The marginal cost of $.00 could stimulate construction of needed distribution facilities since the actual cost of the water to the grower might be very low in this case.

Bruvold, Olson, and Rigby, "Public Policy," 101.

[47] EPA, *EPA's Policy Promoting the Beneficial Use of Sewage Sludge and the New Proposed Technical Sludge Regulations,* Office of Municipal Pollution Control (1989), 1.

[48] EPA, *EPA's Policy Promoting the Beneficial Use of Sewage Sludge,* 6.

[49] WPCF, *CWA 1987,* 9–10.

[50] H. W. Bryan and C. J. Lance, "Compost Trials on Vegetables and Tropical Crops," *Biocycle* (March 1991), 36–37.

[51] WPCF, *CWA 1987,* 6.

[52] EPA, *EPA's Policy Promoting the Beneficial Use of Sewage Sludge,* 7.

[53] EPA, *EPA's Policy Promoting the Beneficial Use of Sewage Sludge,* 8.

[54] EPA, *Data Sheet on Sludge Biosolids Product Values* (undated).

[55] EPA, *Environmental Investments, The Cost of a Clean Environment* (1990), 2-2 to 2-3 (Table 2-1).

[56] Center for Resource Economics, *Analysis of Environmental Protection Funding,* Report to the House-Senate Conference on the 1993 VA, HUD, Independent Agencies Appropriations Act (1992), 11.

[57] Letter from Roberta H. Savage, Executive Director, Association of State and Interstate Water Pollution Control Administrators, to Honorable Max Baucus, Chair, Senate Environment Subcommittee (Dec. 23, 1991), 2; Statement of Bill Frank, Jr., Chair, Northwest Indian Fisheries Commission, before the House VA, HUD, Independent Agencies Appropriations Subcommittee (May 1, 1991), 1.

[58] EPA, *1990 Needs Survey Report to Congress,* Assessment of needed publicly owned wastewater treatment facilities in the United States–including federally recognized Indian tribes and Alaska native villages, Office of Water, EPA-430/09-91-024 (1991), 1; Testimony of Roberta H. Savage, President, America's Clean Water Foundation and Executive Director, Association of State and Interstate Water Pollution Control Administrators, before the Senate Committee on Environment and Public Works (Sept. 22, 1992).

[59] Jonathan C. Kaledin, ed., *Cause for Concern: America's Clean Water Funding Crisis* (National Water Education Council, 1992), 12.

[60] Kaledin, *Cause for Concern,* 3, 29.

[61] EPA, *Environmental Investments: The Cost of a Clean Environment—A Summary,* Office of Policy, Planning, and Evaluation, EPA-230/12-90-984 (1990), 4-4 to 4-5.

[62] National Association of Flood and Stormwater Management Agencies, *Municipal Separate Storm Sewer System Permit Application Costs* (1992), 1, 5.

[63] EPA, *Cost of a Clean Environment,* vii, 2–6.

[64] National Utility Contractors Association, *A Report on Clean Water Investment and Job Creation,* Apogee Research (1992), 6.

[65] Memorandum from Dennis King, Associate Director, Maryland International Institute for Ecological Economics, to William Painter, Chief, Water Policy Branch, Office of Policy, Planning and Evaluation, EPA (Nov. 30, 1992).

[66] NRC, *Restoration of Aquatic Ecosystems: Science, Technology, and Public Policy* (Washington, D.C.: National Academy Press, 1992).

[67] Mark H. Dorfman, Warren R. Muir, and Catherine G. Miller, *Environmental Dividends: Cutting More Chemical Wastes,* INFORM (1992), 11, 19–20.

[68] Greg Karras, *Clean Safe Jobs, The Benefits of Toxic Pollution Prevention and*

Industrial Efficiency to the Communities of South San Francisco Bay (CBE, 1992), iv, 4, 8, 10, 12–15.

[69] NRC, *Alternative Agriculture* (Washington, D.C.: National Academy Press, 1989), 3–6.

CHAPTER 4

[1] CWA § 101(a)(2). States are not permitted to establish uses such as "waste assimilation" but must classify all waters as "fishable and swimmable" unless they meet stringent standards of proof that these uses are not achievable. *See* 40 CFR § 131.10(g).

[2] 40 CFR § 131.6.

[3] 40 CFR pt. 131.

[4] CWA § 303(c)(3), (c)(4).

[5] 40 CFR § 131.12. While EPA considers antidegradation to be technically a part of the water quality standards themselves, it is, for practical purposes, a program designed to implement water quality standards in a way that ensures that clean waters remain clean.

[6] Within one year after enactment of the act, EPA was required to develop and publish water quality criteria

> accurately reflecting the latest scientific knowledge (A) on the kind and extent of *all identifiable effects on health and welfare* including, but not limited to, plankton, fish, shellfish, wildlife, plant life, shorelines, beaches, esthetics, and recreation which may be expected from the presence of pollutants *in any body of water, including ground water,* (B) *on the concentration and dispersal of pollutants, or their byproducts, through biological, physical, and chemical processes;* and (C) on the effects of pollutants on biological community diversity, productivity, and stability, including information on the factors affecting rates of eutrophication and rates of organic and inorganic sedimentation for varying types of receiving waters.

CWA § 304(a)(1), emphasis added.

At the same time, EPA was to publish

> information (A) on the factors necessary to restore and maintain the chemical, physical, and biological integrity of all navigable waters, ground waters, waters of the contiguous zone, and the oceans; (B) on the factors necessary for the protection and propagation of shellfish,

fish, and wildlife for classes and categories of receiving waters and to allow recreation in and on the water.

CWA, § 304(a)(2).

7   NRDC et al. v. Train, 8 ERC 2120 (D.D.C. 1976), *modified,* 12 ERC 1833 (D.D.C. 1979). Paragraph 11 of the consent decree provides, in relevant part:

> The Administrator shall publish, under Section 304(a) of the Act, water quality criteria accurately reflecting the latest scientific knowledge on the kind and extent of all identifiable effects on aquatic organisms and human health of each of the pollutants listed in Appendix A. Such water quality criteria shall state, *inter alia,* for each of the pollutants listed in Appendix A, the recommended maximum permissible concentrations (*including where appropriate zero*) consistent with the protection of aquatic organisms, human health and recreational activities.

> 12 ERC 1843 (as modified), emphasis added.

8   GAO, "Water Pollution: Stronger Efforts Needed by EPA to Control Toxic Water Pollution," GAO/RCED-91-154 (1991), 29.

9   GAO, "Water Pollution: Stronger Efforts Needed," 29.

10   GAO, "Water Pollution: Stronger Efforts Needed," 17.

11   Sheiman, *The Right to Know More,* 8–9.

12   EPA, *1990 TRI, Public Data Release,* EPA-700/S-92-002 (1992), 187–89.

13   GAO, "Water Pollution: Stronger Efforts Needed," 29.

14   EPA, *Integrated Risk Assessment for Dioxins and Furans from Chlorine Bleaching in Pulp and Paper Mills,* EPA-560/5-90-011 (1990), 15, 69. This figure was based on an observed effect level at 0.038 ng/L, with a factor of 1,000 to account for acute v. chronic exposure, differences in species' sensitivities, and differences in field v. laboratory effects. No safety factor was added. Id., 15.

15   Mehrle et al., "Toxicity and Bioconcentration of 2,3,7,8-Tetrachlorodibenzo-dioxin and 2,3,7,8-Tetrachlorodibenzofuran in Rainbow Trout," *Environmental Toxicology and Chemistry* 7, no 1 (1988), 47–62.

16   EPA, *Integrated Risk Assessment for Dioxins,* 34–37.

17   Foran, *Regulating Toxic Substances,* 58.

18   EPA, *Ambient Water Quality Criteria for 2,3,7,8-Tetrachloro-dibenzo-p-diox in,* EPA-440/5-84-007 (1984), C-14.

19   EPA, *Integrated Risk Assessment for Dioxins,* 35; EPA, *Assessing Human Health*

Risks from Chemically Contaminated Fish and Shellfish: A Guidance Manual, EPA-503/8-89-002 (1989), App. F.

20 Foran, *Regulating Toxic Substances,* 58–59.

21 This comment is made with some reservation, as flowing rivers simply transfer pollutants downstream to lakes, estuaries, and marine systems. Nevertheless, as shown by experience with the Great Lakes, systems with high residence times can accumulate dangerous concentrations of toxics in water, sediment, and biota. *See* Foran, *Regulating Toxic Substances,* 81.

22 Returning again to the dioxin example, EPA's integrated risk assessment noted that 2,3,7,8-TCDD in effluent from chlorine-bleaching pulp and paper mills "could be exerting significant adverse effects on aquatic life and on avian and mammalian predators feeding upon aquatic life." EPA, *Integrated Risk Assessment for Dioxins,* 70. Yet no numeric criteria have been issued to address these risks.

23 EPA and FWS, U.S. Dept. of Interior, *1982 National Fisheries Survey,* vol. 1 Technical Report: Initial Findings (1984), iii.

24 EPA, *Feasibility Report on Environmental Indicators,* 82.

25 EPA, *Biological Criteria: National Program Guidance for Surface Waters.*

26 EPA, *Environmental Indicators for Surface Water Quality Programs: Pilot Study,* 4–17.

27 EPA, *Feasibility Report on Environmental Indicators,* 83.

28 CWA § 303(c)(2)(B), added by the Water Quality Act § 308(d).

29 A number of reasons for the states' reluctance to adopt numerical criteria have been cited, including a lack of confidence in the accuracy and currentness of the EPA's data and a fear that they will be subject to legal challenges. Thus the perceived and actual deficiencies in EPA's criteria documents have contributed to state inaction. GAO, "Water Pollution: Stronger Effort Needed," 30.

30 EPA, *National Pretreatment Program: Report to Congress,* Office of Water, 21W-4004 (1991), 6–47.

31 "State Compliance with Clean Water Act Requirements for Adoption of Water Quality Criteria for Toxic Pollutants," 55 Fed. Reg. 14350 (April 17, 1990).

32 "Amendments to the Water Quality Standards Regulation To Establish the Numeric Criteria for Priority Toxic Pollutants Necessary to Bring All States into Compliance with Section 303(c)(2)(B)," 56 Fed. Reg. 58420 (Nov. 19, 1991).

33 Under the CWA, once EPA identifies gaps in state standards, it must act within 90 days to correct those deficiencies. CWA § 303(c)(4). After placing the states on formal notice in Nov. 1991, EPA had until Feb. 1992 to act. Based on the

agency's failure to carry out this duty, NRDC filed suit on June 8, 1992. NRDC, Inc., v. Reilly, (D.C. N.J., filed June 8, 1992). NRDC and N.J. PIRG filed a motion for summary judgment on Aug. 21, 1992.

[34] 57 Fed. Reg. 60848 et seq. (Dec. 22, 1992).

[35] Foran, *Regulating Toxic Substances,* 92–96.

[36] ESA § 7(a)(2).

[37] Details of this dispute can be found in the pleadings and briefs in Mudd v. Reilly, no. CV-91-P-1392-S (N.D.Ala.).

[38] Memorandum of Understanding, coordination between the EPA, FWS, and NMFS regarding development of water quality criteria and water quality standards under the Clean Water Act (July 27, 1992).

[39] GAO, "Water Pollution: Stronger Efforts Needed," 22–23, citing EPA, *National Water Quality Inventory, 1988.*

[40] GAO, "Water Pollution: Stronger Efforts Needed," 24–25.

[41] GAO, "Water Pollution: Stronger Efforts Needed," 22.

[42] The rate of monitoring for toxic pollutants appears to be increasing. The 1990 *Water Quality Inventory* found that 41 states reported that they had done some monitoring for toxic substances in 182,611 river miles (out of a total of 1.8 million stream miles identified in the report), as compared to only 28 states in 1988 having reported that they monitored for toxics in about 67,000 river miles. EPA, *National Water Quality Inventory, 1990* (April 1992), 86, 3; *National Water Quality Inventory, 1988* (April 1990), 102. Even this increased rate of toxics monitoring, however, means that only about 10% of river miles have been monitored to any degree.

[43] GAO, "Water Pollution: Stronger Efforts Needed," 24.

[44] EPA requires states to have "appropriate monitoring methods and procedures... necessary to compile and analyze data on the quality of waters of the United States." 40 CFR § 130.4(a). This program must "include collection and analysis of physical, chemical and biological data and quality assurance and control programs to assure scientifically valid data." Id., § 130.4(b).

[45] GAO, "Water Pollution: Stronger Efforts Needed," 26–27.

[46] EPA, *Environmental Indicators for Surface Water Quality Programs,* E-6, E-9, 4-1.

[47] NRC, *Managing Troubled Waters: The Role of Marine Environmental Monitoring* (Washington, D.C.: National Academy Press, 1990), *cited in* O'Connor, *Coastal Environmental Quality,* 1.

[48] EPA, *Feasibility Report on Environmental Indicators,* 105–6.

[49] EPA, *National Pretreatment Program,* 6–47.

[50] CBF, *Chesapeake Bay Sewage Treatment Plant Performance Review* (1991), 9–10. These included 35 in Md., 21 in Va., and 3 in Pa., less than half of the plants surveyed. Id.

[51] For example, *Surface Water Monitoring: A Framework for Change;* EPA *Feasibility Report on Environmental Indicators* (1990); Aug. 1991 Guidelines for 305(b) reports; *Environmental Indicators for Surface Water Quality Programs* (1992).

[52] Intergovernmental Task Force on Monitoring Water Quality, *Ambient Water-Quality Monitoring in the United States: First Year Review, Evaluation, and Recommendations* (EPA, 1992).

[53] NRDC, *Testing the Waters*, 2. States that monitor their beach waters infrequently if ever include Ala., Ga., La., Miss., N.H., N.C., Oreg., S.C., Tex., and Wash.; states that monitor portions of their coasts are Calif., Del., Fla., Maine, Md., Mass., R.I., and Va.; states that monitor their whole coasts are Conn., N.J., N.Y., and Del. Id., 15.

[54] NRDC, *Testing the Waters*, 15–16. EPA has responded to the concerns about inadequate monitoring, testing procedures, and closure/advisory practices revealed in *Testing the Waters* by considering a plan to convene a group of interested stakeholders to undertake a negotiated rule-making. The purpose of the negotiations would be to seek common agreement regarding the best testing procedures, monitoring frequencies, and closure standard. The negotiations also may include closure/advisory requirements. Mike Hanlon, EPA Office of Water, personal communication (Aug. 1992).

[55] Patricia A. Cunningham, Ph.D., J. Michael McCarthy, and Devorah Zeitlin, *Results of the 1989 Census of State Fish/Shellfish Consumption Advisory Programs*, prepared for EPA Assessment and Watershed Protection Division, Office of Water Regulations and Standards (Research Triangle Institute, 1990), 3.1, 4.1.

[56] Devorah Zeitlin, ed., *State-Issued Fish Consumption Advisories: A National Perspective* (NOAA, 1990), 9.

[57] Zeitlin, *State-Issued Fish Consumption Advisories*, 12.

[58] Foran, *Regulating Toxic Substances*, 63, 66.

CHAPTER 5

[1] CWA defines a *discharge* as "any addition of any pollutant to navigable waters from any point source." CWA § 502(12). Pollution sources that do not come from a "discrete conveyance"–i.e., from a "point" at the end of a pipe–are often referred to as *nonpoint sources.* Discussed in greater detail following,

this category incudes essentially anything that is not a point source, from agricultural runoff to stormwater contamination from a construction site or parking lot.

2   EPA, *De Minimus Discharges Study: Report to Congress,* EPA-440/4-91-002 (1991), 1, 2, 13.

3   CRS, *1972 Legislative History,* 1425.

4   CWA § 101(a)(1).

5   CWA §§ 304(b)(3), 304(c), 306(a)(1).

6   CWA §§ 301(b)(1)(B), 304(d)(1). *Secondary treatment* is the standard for removal or treatment of conventional pollutants–i.e., pollutants associated with domestic sewage–such as suspended solids, biological oxygen demand, and pH. *See* 40 CFR pt. 133. The new secondary treatment standard was coupled with a major federal grants program to finance construction of wastewater treatment facilities around the nation. CWA, Title II.

7   CWA §307(b)(1).

8   CWA § 402(b).

9   CWA §§ 301(a), 402(a)(1).

10   40 CFR pt. 403.

11   CWA § 301(b)(1)(C).

12   CWA §§ 402(a), 301, 306, 307.

13   CWA §§ 301(b)(2)(A), 304(b)(3).

14   CWA § 402(b). Section 402 requires compliance with effluent limits and other applicable standards "or prior to the taking of necessary implementing actions...such actions as the Administrator [or state] determines are necessary to carry out the provisions of the Act."

15   GAO, "Water Pollution: Stronger Efforts Needed," 31. *Also* EPA, *Assessment of State Needs for Technical Assistance in NPDES Permitting,* Program Evaluation Division (1984).

16   Office of Technology Assessment (OTA), U.S. Congress, *Wastes in Marine Environments* (1987), 186–87, citing EPA, "Summary of Effluent Characteristics and Guidelines for Selected Industrial Point Source Categories: Industry Status Sheets," Office of Regulations and Standards, Monitoring and Data Support Division (Feb. 1986).

Priority pollutants are the main focus of the current effluent guidelines, but both EPA and other investigators have concluded that this focus does not go far enough. Some industries discharge large quantities of toxic and haz-ardous pollutants that are not on the priority pollutant list but need to be regulated. *See* EPA, *Report to Congress on the Discharge of Hazardous Wastes to Publicly Owned Treatment Works,* commonly referred to as the "Domestic Sewage Study," or "DSS" (1986), 6-63, 6-64.

[17] EPA, *1990 TRI,* at 15. The *TRI* data collection and reporting effort now con-tains four years' worth of data on industrial releases to all media (air, water, land disposal, public sewers, and off-site disposal). The trends in reported water releases suggest a general downward shift: 1987, 412 million lb/yr; 1988, 311 lb/yr; 1989, 193 lb/yr; and 1990, 197 lb/yr.

However, the downward trend must be read with caution. First, *TRI* data can be based on estimates, and a reporting facility can change the estimating technique over time. Second, a facility with high reported discharges is not prevented from shifting chemicals used in its manufacturing processes to substitute a substance for which no reporting requirement exists. Third, a number of large-volume but relatively low-toxicity chemicals have been delisted since 1988 and are no longer reported.

[18] G. Lomas et al., *Toxic Truth and Consequences: The Magnitude of and the Problems Resulting From America's Use of Toxic Chemicals,* National Environmental Law Center and U.S. PIRG (1991).

[19] OTA, U.S. Congress, statement before the Subcommittee on Superfund, Ocean and Water Protection, Committee on Environment and Public Works, U.S. Senate (May 10, 1989).

A large percentage of reported total toxic releases are to air rather than to water. While these releases are initially airborne, a significant but currently unknown percentage of them wind up in the water, either through direct air-to-water deposition (in the Great Lakes, for example, this deposition route is believed to be highly significant because of the lakes' huge surface area) or through land deposition, after which it is transported to the water through pol-luted runoff.

[20] NAS, *Drilling Discharges in the Marine Environment* (1983), 16.

[21] U.S. Dept. of Interior, *Final Environmental Impact Statement, 5-Year Outer Continental Shelf Oil and Gas Program, 1992–1997,* Table IV.A.1-I.

[22] EPA, *Preliminary Data Summary for the Coastal, Onshore and Stripper Subcategories of the Oil and Gas Extraction Point Source Category,* draft (1989), 10.

[23] According to a recent EPA study, 17,463 direct dischargers are subject to BAT standards. Another 31,958 facilities are in categories that have no national stan-

dards, and 4,031 are of unknown industrial categories but also lack BAT standards. EPA, *De Minimis Discharges Study*, 13. The *1989 Water Quality Improvement Study*, at 17, paints an even bleaker picture, estimating that nearly four-fifths of all facilities remain outside the coverage of national BAT standards. As will be discussed following, the situation is still more extreme when it comes to the establishment of national pretreatment standards applicable to industries discharging their wastewaters to POTWs.

[24] All existing effluent guidelines are set forth in 40 CFR pts. 405–71.

[25] EPA, "Notice of Proposed Effluent Guidelines Plans," 57 Fed. Reg. 19748, 19756-57 (May 7, 1992).

[26] *Effluent Guidelines and Pretreatment Standards for the Organic Chemicals, Plastics and Synthetic Fibers Category*, 40 CFR pt. 414. *Effluent Guidelines and New Source Performance Standards for the Oil and Gas Extraction Category, Offshore Subcategory*, 40 CFR pt. 435.

[27] HR Rep. No. 100-4, 99th Cong., 2d. Sess. 128 (Oct. 15, 1986).

[28] CWA § 304(m), as added by § 308(f) of the WQA of 1987.

[29] 55 Fed. Reg. 80 (Jan. 2, 1990).

[30] Under consent decrees in three cases, EPA will issue guidelines according to the following schedules:

Industry	Promulgation Dates
Offshore oil & gas extraction	1/92
OCPSF revisions	5/93
Pesticide chemicals mfg.	7/93
Pulp, paper, & paperboard	9/95
Pesticide chemicals formulating/packaging	8/95
Waste treatment (Phase 1)	1/96
Pharmaceutical mfg.	2/96
Metal products & machinery (Phase 1)	5/96
Coastal oil & gas extraction	7/96
Waste treatment (Phase 2)	1997
Industrial laundries	1998
Transportation equipment cleaning	1998
Metal products and machinery (Phase 2)	1999
Eight additional categories (see below)	By 2003

EPA will also study the following industries and will promulgate eight additional effluent guidelines based upon these studies:

Industry	Study Completion Date
Petroleum refining	1993
Metal finishing	1993
Iron and steel	1994
Inorganic chemicals	1994
Leather tanning	1995
Coal mining	1996
Onshore/stripper oil and gas	1996
Three additional categories (to be selected by EPA)	1997

NRDC v. Reilly, Civ. No. 89-2980 (D.C.D.C.); NRDC v. Thomas, Civ. No. 79-3442 (D.C.D.C.); EDF v. Thomas, Civ. No. 85-0973 (D.C.D.C.). The Oil and Gas guidelines were issued on Jan. 15, 1993.

[31] CWA § 301(b)(2)(A) (emphasis added); id., § 304(b)(3). EPA must "identify control measures and practices available to eliminate the discharges of pollutants from categories and classes of point sources."

[32] CWA § 304(b)(1)-(2).

[33] Examples include some onshore oil and gas wells, 40 CFR pt. 435, subpt. B, and some placer gold mines, 40 CFR pt. 440, subpt. M. *See* Oliver Houck, "The Regulation of Toxic Pollutants Under the Clean Water Act," *Environmental Law Reporter* 21 (1991), 10528, 10538.

[34] Houck, "Regulation of Toxic Pollutants," 10538–39.

[35] 52 Fed. Reg. 42558-59 (Nov. 5, 1987).

[36] Chemical Manufacturers' Assn. v. EPA, 870 F.2d 177, 261–64 (5th Cir., 1988).

[37] 56 Fed. Reg. 63897 et seq. (Dec. 6, 1991).

[38] EPA, *De Minimis Discharges Study*, 13.

[39] CWA § 301(b)(1)(B).

[40] EPA, *Report of the Strategic Options Subcommittee*, Science Advisory Board, Relative Risk Reduction Project (Sept. 1990), App. C, 10–11.

[41] EPA, *Report to Congress on the Discharge of Hazardous Wastes to Publicly Owned Treatment Works*, EPA-530/SW-86-004 (1986), E-3.

[42] EPA, *Pretreatment Study*, 1–4.

[43] OTA, *Wastes in Marine Environments*, 185.

[44] EPA, *1990 TRI,* 15.

[45] EPA, Decision Document for Martha Prothro, Director, Office of Water Regulations and Standards, on Hazardous Waste Treatment Facilities (undated).

[46] OTA, *Wastes in Marine Environments,* citing sources 40 and 38 therein.

[47] EPA, *Pretreatment Study,* 7-55.

[48] EPA, *Pretreatment Study,* 7-15.

[49] EPA, *Pretreatment Study,* 7-19.

[50] *Statistical Assessment of National Significant Industrial User Noncompliance: Final Report,* submitted to EPA, Office of Wastewater Enforcement and Compliance, by SAIC (June 1992), 2-1. EPA policies define *Significant Industrial Users* and *Significant Noncompliance,* and this study relies upon those official definitions to evaluate the number of facilities that met the definitions. If all violations–significant and those defined as not significant– are included, then the totals are even more appalling: 81% of all Significant Industrial Users violated at least one pretreatment requirement (discharge or reporting) during the study period. Id., 2-2. The study was conducted to obtain baseline data; at the time the violations occurred, in 1990, the definitions had not yet been adopted into the regulations. This in no way affects the significance of the findings, however.

[51] GAO, "Water Pollution: Improved Monitoring and Enforcement Needed for Toxic Pollutants Entering Sewers," GAO/RCED-89-101 (1989), 4.

[52] EPA, "Sewage Sludge Use and Disposal Rule (40 CFR Part 503)–Fact Sheet," EPA-822/F-92-002 (1992).

[53] "EPA Policy on Municipal Sludge Management," 49 Fed. Reg. 24358 (June 12, 1984); "Notice of interagency policy on beneficial use of sewage sludge on Federal land," 56 Fed. Reg. 30448 (July 2, 1991).

[54] EPA, "Sewage Sludge Use and Disposal Rule."

[55] GAO, "Water Pollution: Nonindustrial Wastewater Pollution Can Be Better Managed," GAO/RCED-92-40 (1991), 3.

[56] GAO, "Water Pollution: Nonindustrial Wastewater," 13–16.

[57] EPA, "Guidance for Writing Case-By-Case Permit Requirements for Municipal Sewage Sludge," EPA-505/8-90-001 (1990), Table E-6, 185 et seq.

[58] N.Y. state recently has encountered such resistance to its proposed export of NYC sludge that does not meet the relatively strict N.Y. state criteria. The state sought land application sites in Tex. and Okla., both of which lack comprehensive state sludge quality standards; the move prompted legal challenges in both states.

[59] CWA § 405(d).

[60] NRDC v. EPA, 790 F.2d 289 (3d Cir. 1986).

[61] CWA § 405(d)(2)(A) and (B), respectively.

[62] CWA § 405(f)(1). This provision actually requires that the sludge management regulations be established by a date that antedates passage of the 1987 WQA Amendments, an artifact of the length of time it took Congress to enact the law.

[63] CWA § 405(f).

[64] 40 CFR pt. 501; see 54 Fed. Reg. 18786 (May 2, 1989).

[65] 40 CFR pt. 501; see 54 Fed. Reg. 18786 (May 2, 1989).

[66] As of Jan. 1993, not a single state had assumed delegated authority to manage its sludge program, nor had any formal applications for delegated authority been filed with the agency; while several states had met with EPA headquarters or regional staff regarding delegation, no action had yet been taken. Pamela Mazakas, telephone conversation, EPA Office of Water, Office of Wastewater Compliance and Enforcement, Permits Division, Pretreatment and Multi-Media Branch (Jan. 11, 1993). The states expressing an interest in delegated authority are Wis., W.Va., and N.J.

[67] "Solid Waste Disposal Facility Criteria," 56 Fed. Reg. 50978 (Oct. 9, 1991), codified largely as 40 CFR pt. 258. EPA's failure to issue these and the other set of sludge quality regulations described in the text was the subject of several "deadline suits" under the CWA and the Administrative Procedure Act. See NRDC v. Reilly, Civ. No. 88-2101 (E.D. Pa.; dismissed Nov. 5, 1991); Gearhart v. Reilly, Civ. No. 89-6266-HO (D.Ore.).

[68] "Standards for the Use or Disposal of Sewage Sludge," 58 Fed. Reg. 9248. (Feb. 19, 1993).

[69] CWA § 402(d)(3).

[70] CWA § 307(b), which establishes the pretreatment program, authorizes the removal credit program on the theory that industry should not be required to conduct "redundant" treatment of substances that are adequately treated by the POTW itself. The law specifically limits removal credits to cases where the increased toxic discharge will not "prevent sludge use or disposal" by the POTW. Id. Of course, "removal" of a pollutant from the water effluent to the sludge hardly constitutes "treatment."

[71] "EPA Policy on Municipal Sludge Management," 49 Fed. Reg. 24358 (June 12, 1984).

[72] Sierra Club v. EPA, No. 92-1003, D.C. Cir, argued Jan. 11, 1993. Just before this book went to press, the court rejected most of the challenges to these rules.

[73] EPA received 5,500 pages of comments from 630 commenters. Some argued that the proposed rules were inadequately protective; many argued that they overstated sludge's risk and would be overly stringent.

[74] CWA § 405(d)(2)(D).

[75] CWA § 402(f).

[76] 40 CFR pts. 257, 403 and 503 (FRL-4203-3), "Standards for the Use and Disposal of Sewage Sludge: Final Rule"; "Overview" section, "Fundamental Regulatory Principles: Promulgate an Implementable Rule," rule available only on computer disk at press time.

[77] GAO, "Water Pollution: Serious Problems Confront Emerging Municipal Sludge Management Program," GAO/RCED-90-57 (1990). As early as this report, the GAO flagged concerns about inadequate resources for implementation–a problem GAO called a "generic" one affecting environmental programs at EPA. Id., 25–27. GAO also questioned EPA's readiness to carry out adequate enforcement and provide adequate headquarters oversight of state programs at EPA. Id., 28–32. Finally, GAO noted that coordination problems might arise if differing federal and state regulatory authorities controlled sludge and discharge permitting–something likely to occur under the sludge management rules EPA promulgated. Id., 35–36.

[78] 40 CFR §§ 503.13, 503.23, and 503.40.

[79] CWA § 405(d)(2)(C).

[80] EPA, *De Minimis Discharges Study*, 1–2.

[81] OTA, *Wastes in Marine Environments*, 191.

[82] EPA, *Inspector General Special Review Report*, E1HWG9-13-9024-0400018 (June 14, 1990).

[83] 55 Fed. Reg. 47991 (1990).

[84] 55 Fed. Reg. 47991. (Although NURP did not attempt to study oil and grease in urban runoff, EPA cites other studies that documented significant quantities of oil and grease in urban storm water.)

[85] 55 Fed. Reg. 47992. (1990).

[86] NRDC, "Poison Runoff Indexes for Washington, D.C. (November 1989); Baltimore, MD (November 1989); Tidewater, VA (February 1991); Harrisburg, PA (September 1990); and Los Angeles, CA (October 1990)."

[87] 55 Fed. Reg. 47991 (1990).

[88] 55 Fed. Reg. 47992 (1990).

[89] EPA, Draft Report to Congress under CWA § 402(p)(5): *The Identification,*

NOTES

Nature and Extent of Stormwater Discharges in the United States, Table B-16.

[90] 38 Fed. Reg. 1350 (May 22, 1973).

[91] NRDC v. Train, 396 F. Supp. 1393 (D.D.C. 1975; Aff'd, NRDC v. Costle, 568 F.2d 1369, D.C. Cir. 1977). The Court held that Congress intended all point source discharges to be subject to the NPDES permit program. Id., 1396. The Court reasoned that "[t]o allow the exemptions made by the Administrator is to diminish the effect of the Act.... If a point source is exempted from the permit requirement, the Administrator then has no effective control over the polluter." Id., 1399. The court acknowledged that EPA had been assigned expansive tasks but nevertheless ruled that "[t]he compelling congressional intent is clearly present. It is expressed in the statute itself and in the legislative history, both of which demonstrate that the discharge of pollutants without a permit is unlawful." Id., 1400.

[92] The details of this rule-making history are set out in NRDC's Opening Brief in NRDC v. EPA, Nos. 90-70671 and 91-70200 (9th Cir., U.S. Court of Appeals).

[93] CRS, *A Legislative History of the Water Quality Act of 1987*, comm. print no. 1, 100th Cong., 2d Sess. 1304, 617 (hereafter cited as *1987 Legislative History*) (1988). Senators Mitchell and Chafee noted:

> Runoff from municipal separate storm sewers and industrial sites contains significant volumes of both toxic and conventional pollutants, including 13 toxic metals, in the discharge from municipal separate storm sewers that were studied. Of these, lead, copper, and zinc were the most pervasive; EPA found these pollutants in at least 91 percent of its samples. The same study also estimated that municipal separate storm sewers discharge 10 times the total suspended solids that the Nation's secondary sewage treatment plants discharge.
>
> Toxic and conventional storm water contaminants may adversely affect public health, harm fish and other aquatic life, and prevent or retard water quality improvements even when the best available pollution controls are installed on other point sources.

Id., 391, 646.

[94] In § 405 of the 1987 WQA (adding § 402(p) to the CWA), Congress established explicit and firm deadlines for EPA regulation of stormwater discharges. Section 402(p)(1) of the act provides that EPA cannot require a permit for certain stormwater discharges until Oct. 1, 1992, with five exemptions (discharges that are required to obtain a NPDES permit prior to Oct. 1, 1992): (1) a discharge with respect to which a permit has been issued prior to February 4, 1987; (2) a discharge associated with industrial activity; (3) a discharge from a munici-

pal separate storm system serving a population of 250,000 or more; (4) a discharge from a municipal separate storm sewer system serving a population of 100,000 or more, but less than 250,000; or (5) a discharge for which the administrator or the state determines that the stormwater discharge contributes to a violation of a water quality standard or is a significant contributor of pollutants to the waters of the United States. The statute makes clear that all stormwater discharges associated with industrial activities remain subject to all requirements of §§ 301 and 402 of the act but subjected discharges from municipal storm sewers to new requirements. CWA § 402(p)(3)(B).

[95] In late 1992, Congress once again granted EPA and these stormwater sources a reprieve, giving EPA until Oct. 1, 1993, to issue regulations and giving the sources until Oct. 1, 1994, to submit permit applications.

[96] EPA, *1988 Needs Survey Report to Congress: Assessment of Needed Publicly Owned Wastewater Treatment Facilities in the United States,* Office of Municipal Pollution Control (1989), 1, 15.

[97] EPA, *1988 Needs Survey Report,* 15.

[98] EPA, *1988 Needs Survey Report,* 12. EPA's documentation requirements are detailed in the *Needs Survey,* at App. D. These requirements (which effectively keep out of the official count those needs that do not meet the requirements) have the effect of underestimating the CSO problems that the states themselves deem to be in need of correction. The states of Ill., Maine, N.J., N.Y., Oreg., Pa., and Wash. estimated that they had an additional "separate" (i.e., not "documented" in accordance with EPA requirements) need for CSO correction funds of nearly $2.14 billion in 1988. Id., App. A-7.

[99] EPA, *1988 Needs Survey Report,* 15. The same pattern holds true for "separate" needs. Six of the seven states citing additional separate CSO needs outside the documented needs in the *Needs Survey* are marine coastal states; the other is a Great Lakes state. Id., App. A-7.

[100] EPA, *1988 Needs Survey Report,* 12.

[101] 1990 U.S. population was estimated to be 249 million. NOAA, *Fifty Years of Population Change along the Nation's Coasts, 1960–2010* (1990), 4.

[102] EPA, "Seminar Publication: Benefit Analysis for Combined Sewer Overflow Control," Office of Technology Transfer (undated), 2, citing A. Lager et al., "Urban Stormwater Management and Technology, Users Guide" (1977).

[103] EPA, "Benefit Analysis," 2.

[104] 40 CFR § 133.102.

[105] EPA, "Benefit Analysis," 2.

[106] *See,* for example, the State Water Quality Standards for Fla., Md., N.Y., and N.C. NRDC, *Testing the Waters,* 12.

[107] For an explanation of this derivation, see Chapter 2, note 10.

[108] EPA, *Report to Congress on the Discharge of Hazardous Wastes to Publicly Owned Treatment Works* (1986), E-3.

[109] EPA, Status of Combined Sewer Overflows Strategy Approvals (current as of July 20, 1992).

[110] EPA, *Combined Sewer Overflow Control Policy,* draft (Dec. 18, 1992).

[111] These include no more than four overflows per year in urban areas and no more than five in rural areas; or capture or elimination of 85% of overflows by volume or the equivalent in pollutant mass. EPA, *Combined Sewer Overflow Control Policy,* 18–19. Cities may adopt alternative controls if they can demonstrate compliance with water quality standards and protection of designated uses.

[112] EPA, "National Pollutant Discharge Elimination System; Surface Water Toxics Control Program," proposed rule; "Surface Water Toxics Control Program and Water Quality Planning and Management Program," final rule, 57 Fed. Reg. 33040, 33054 (July 24, 1992).

[113] CWA § 402(o).

[114] CBF, *Chesapeake Sewage Treatment Plant.* CBF surveyed 160 treatment plants in the Chesapeake Bay watershed, which remains severely impaired by nutrients, toxics, and other pollutants (47 in Md., 83 in Pa., and 30 in Va.). Only Md. required total N limits, and for only three plants. All states require phosphorus limits, but not for all plants (53% of plants in Md., 42% in Pa., 33% in Va.). None in Md. or Va. regulated priority toxics; only 1 did in Pa. (cyanide). 35 in Md., 21 in Va., and 3 in Pa. had permit requirements for biomonitoring (whole effluent toxicity).

[115] 57 Fed. Reg. 33040, 33054 (July 24, 1992).

[116] CRS, *1972 Legislative History,* 166. Similarly, the Senate Report declared that the "use of any river, lake, stream or ocean as a waste treatment system is unacceptable." Id., 1425.

[117] Mark C. Van Putten and Bradley D. Jackson, "The Dilution of the Clean Water Act," *Journal of Law Reform* 19, no. 4 (Summer 1986), 863–901.

[118] CWA § 303(d)(1)(C).

[119] CWA § 303(c)-(e). The details of this process are set forth in EPA regulations. 40 CFR pt. 130.

[120] *See generally* Foran, *Regulating Toxic Substances,* 74–83.

[121] GAO, *Water Pollution: More EPA Action Needed to Improve the Quality of Heavily Polluted Waters,* GAO/RCED-89-38 (1989), 4.

[122] 33 USC § 1314(l).

[123] Based on a relatively quick review of state files, NRDC identified roughly twice as many polluted waters as the states of Ga. and Va. had reported. In response to petitions filed by NRDC, EPA added roughly half of these waters to the state lists. NRDC also filed petitions covering all 50 states, noting serious limitations in state and EPA review procedures.

[124] NRDC v. EPA, 915 F.2d 1314 (1990).

[125] 57 Fed. Reg. 33051 et seq. (July 24, 1992).

[126] Houck, "Regulation of Toxic Pollutants," 10545.

[127] 40 CFR § 131.13.

[128] EPA, *Water Quality Standards Handbook,* Office of Water Regulations and Standards (1983), 2–8.

[129] EPA, *Technical Support Document for Water Quality-based Toxics Control,* Office of Water, EPA/505/2-90-001 (1991), 33–34.

[130] The examples are drawn primarily from Foran, *Regulating Toxic Substances,* 78–83.

[131] Jeffrey A. Foran, "The Control of Discharges of Toxic Pollutants into the Great Lakes and their Tributaries: Development of Benchmarks," a report to the IJC (George Washington Univ., 1991), xi; Foran, *Regulating Toxic Substances,* 85–101. The study consists of calculations regarding the loadings from a hypothetical discharger into waters of each of the Great Lakes states of a series of toxic pollutants, based on the states' applicable water quality criteria and other elements of their applicable water quality standard such as mixing zones, dilution assumptions, etc.

[132] Foran, "Control of Discharges of Toxic Pollutants"; Foran, *Regulating Toxic Substance,* 92.

[133] Foran, "Control of Discharges of Toxic Pollutants," 9–10.

[134] Sen. Muskie, for example, explained that "the intent is that effluent limitations...within a given category or class be as uniform as possible...to assure that similar point sources with similar characteristics, regardless of their location or the nature of the water in which the discharge is made, will meet similar effluent limitations." CRS, *1972 Legislative History,* 172.

[135] EPA, *Assessment of State Needs for Technical Assistance in NPDES Permitting* (1984).

[136] 57 Fed. Reg. 12560, 12570 (April 10, 1992).

[137] *See* 40 CFR § 122.44.

[138] 40 C.F.R. § 122.45(g).

[139] An example is the iron and steel effluent guideline. 40 CFR, pt 420.

[140] The EPA rule requires only a "reasonable measure of actual production of the facility" for existing sources and "projected production" for new sources. 40 CFR § 122.45(b)(2)(i). Notably, the permit-writer also may establish "alternate" limits based on "anticipated increased...or decreased production levels." Id., § 122.45(b)(2)(ii)(A)(1). Thus, an astute permittee can build in relief in case production levels increase.

[141] Carpenter Environmental Associates, "Review of Ravenswood Aluminum Corporation, West Virginia NPDES Permit Application," prepared for the United Steelworkers of America (1992).

[142] Comments submitted by the NRDC and the Southern Environmental Law Center on Weyerhaeuser NPDES Permit No. NC0003191 (April 14, 1989).

[143] An example of this type of effluent guideline is the OCPSF rule discussed previously. 40 CFR pt. 414.

[144] 40 CFR § 122.45(h).

[145] CWA § 402(b).

[146] CWA § 402(b)(8).

[147] EPA, *De Minimis Discharge Study,* at App. K, p. 1. The non-complying states are Colo., Del., Ill., Ind., Kans., Mont., Nev., N.Y., N.D., Pa., Va., and Wyo.

[148] Gary Hudiburgh, Chief, EPA Regulatory Implementation Section, letter and memorandum to Jessica Landman, NRDC (Feb. 3, 1993). The states in which environmental groups have raised concerns are Kans., Mich., Va., W.Va., Mont., and Wash. The two jurisdictions in which EPA has taken action are Oreg. and the U.S. V.I.

[149] U.S. PIRG and N.J. PIRG, *Permit to Pollute: Violations of the Clean Water Act by the Nation's Largest Facilities* (1991), 4.

[150] U.S. PIRG and N.J. PIRG, *Permit to Pollute,* 4.

[151] Horton and Eichbaum, *Turning the Tide,* 71.

[152] CBF, *Chesapeake Sewage Treatment Plant,* 11.

[153] U.S. PIRG and N.J. PIRG, *Permit to Pollute,* 4–5.

[154] GAO, *Water Pollution: Stronger Efforts Needed,* 19–20.

[155] U.S. PIRG and N.J. PIRG, *Permit to Pollute,* 6.

[156] This was part of a more comprehensive review of criminal enforcement of a

number of environmental laws by EPA and the Dept. of Justice. Robert W. Adler and Charles Lord, "Environmental Crimes: Raising the Stakes," *George Washington Law Review* 59 (1991), 781.

157 Adler and Lord, "Environmental Crimes," 802.

158 Adler and Lord, "Environmental Crimes," App. E.

159 Adler and Lord, "Environmental Crimes," 803–5.

160 EPA, "Review of Region IV's National Pollutant Discharge Elimination System Permit Enforcement Program," Audit Report EIHWD8-04-0207-9100462 (Sept. 7, 1989) (hereafter cited as "Region IV Audit Report"); "Report of Audit on the Management of the Chesapeake Bay Program Point Source Pollution Program," Audit Report E1H98-03-0208-9100467 (September 11, 1989) (hereafter cited as "Chesapeake Bay Audit Report.")

161 EPA, "Region IV Audit Report," 2–5.

162 EPA, "Chesapeake Bay Audit Report," 4–5.

163 EPA, *Inspector General's Report* (June 14, 1990).

164 CBF v. Gwaltney of Smithfield, Ltd., 484 U.S. 49 (1987).

165 A more detailed recitation of enforcement problems is included in the Clean Water Network's *Issue Paper on Enforcement and Accountability.*

CHAPTER 6

1 EPA, *A Report to the Congress: Activities and Programs Implemented Under Section 319 of the Clean Water Act–Fiscal Year 1988* (1989), 7.

2 CRS, *1972 Legislative History,* 1457.

3 EPA did not require states to provide groundwater data in their 319 assessments, so these figures are gross underestimates.

4 Water Quality 2000, *Phase II: Problem Identification, Workgroup Reports* (Sept. 1990), 2.

5 CWA § 304(l); U.S. EPA, "Reporting Status and Quantitative Analysis of 304(l) Lists and Individual Control Strategies" (Aug. 4, 1989).

6 EPA, *Managing Nonpoint Source Pollution: Final Report to Congress on Section 319 of the Clean Water Act (1989)*, EPA-506/9-90 (1992), 17.

7 Statement by Martha Prothro, Deputy Assistant Administrator for Water, EPA, in response to reporter's question. EPA briefing on the 304(l) Toxic Hotspots List (June 13, 1989).

8 EPA, *National Water Quality Inventory: 1990,* (1992), 9.

[9] EPA, *Pesticides and Ground-Water Strategy: A Survey of Potential Impacts,* Office of Pesticide Programs, Biological and Economic Analysis Division (1991), 30.

[10] EPA, *Fish Kills Caused by Pollution, 1977–1987: Summary of Findings 1977–1985* (undated), I-1 to I-2, II-1.

[11] R. P. Maas et al., *Best Management Practices for Agricultural Nonpoint Source Control–IV. Pesticides,* USDA National Water Quality Evaluation Project, Biological and Agricultural Engineering Department, North Carolina State University, USDA Cooperative Agreement 12-05-300-472, EPA Interagency Agreement AD-12-F-O-037-0 (1984), cited in *1986 Water Quality Implications of Conservation Tillage: A Reference Guide* (Ft. Wayne, Ind.,: Conservation Tillage Information Center, 1986).

[12] Anthony S. Pait et al., eds., *Agricultural Pesticides in Coastal Areas: A National Summary,* review copy (NOAA, 1992), 4.

[13] EPA, *National Water Quality Inventory, 1990 Report,* 7, 38.

[14] Thurman et al., "Herbicides in Surface Waters of the Midwestern United States: The Effect of Spring Flush," *Environmental Science and Technology* 25 (1991), 1794–96.

[15] Thurman et al., "Herbicides in Surface Waters" 1794.

[16] J. Fedkiw, *Nitrate Occurrence in U.S. Waters (And Related Questions),* a reference summary of published sources from an agricultural perspective, U.S. Dept. of Agriculture, USDA Working Group on Water Quality (1991), 7.

[17] Fedkiw, *Nitrate Occurrence in U.S. Waters,* 19, 21.

[18] Fedkiw, *Nitrate Occurrence in U.S. Waters,* 21–22.

[19] Fedkiw, *Nitrate Occurrence in U.S. Waters,* 3.

[20] Fedkiw, *Nitrate Occurrence in U.S. Waters,* 7.

[21] Fish and Wildlife Service, *An Overview of Irrigation Drainwater Techniques, Impacts on Fish and Wildlife Resources, and Management Options* (May 1992), i, ii, 14.

[22] FWS, U.S. Dept. of Interior, *An Overview of Irrigation Drainwater Techniques, Impacts on Fish and Wildlife Resources, and Management Options* (1992), iv.

[23] GAO, *Public Rangelands: Some Riparian Areas Restored but Widespread Improvement Will Be Slow,* GAO/RCED-88-105 (1988), 37.

[24] Lynn Jacobs, *Waste of the West: Public Lands Ranching* (1991), 92.

[25] Jacobs, *Waste of the West,* 92.

[26] Jacobs, *Waste of the West,* 22–23.

[27] GAO, *Public Rangelands,* 51–52.

[28] Catherine Long, U.S. EPA, Office of Policy, Planning and Evaluation, personal communication (April 23, 1991). *See* 40 CFR § 122.23 pt. 122, App. B.

[29] Lamonte Garber, ed., "Improving Water Quality Through Effective Implementation of Pennsylvania's Manure Management Regulations" (CBF, Sept. 1989), 3.

[30] Garber, "Improving Water Quality," 1.

[31] EPA, *National Water Quality Inventory, 1990,* 12.

[32] GAO, *Water Pollution: Greater EPA Leadership Needed to Reduce Nonpoint Source Pollution,* GAO/RCED-91-10 (1990), 22.

[33] GAO, *Water Pollution: Greater EPA Leadership Needed to Reduce Nonpoint Source Pollution,* 22.

[34] EPA, *National Water Quality Inventory, 1990,* 12–13.

[35] EPA, *National Water Quality Inventory, 1990,* 12–13.

[36] EPA, "Silviculture," in *Managing Nonpoint Source Pollution, 1992,* 19. The 1988 Oreg. Deq. Report, entitled "1988 Oregon Statewide Assessment of Nonpoint Sources of Water Pollution," does allow the interested reader to compile watershed-based (and component water-body–based) data on land use sources contributing to impairments; thus, EPA's statement about Oreg. not reporting data on silvicultural water quality effects is apparently an error.

[37] Robert Steelquist, ed., "Watershed Wars: Salmon and Forests, Fog Brothers," in *American Forests* (American Forestry Association, July/Aug. 1992), 31.

[38] Maine Department of Environmental Protection, *Nonpoint Source Pollution Assessment and Management Program* (Draft) (April 1, 1988), 24-25.

[39] § 101(e) contains a broad mandate for public participation that has been grossly underemployed in the campaign to stem the flow of runoff. Without widespread public participation in the form of volunteer water quality monitoring programs and citizen involvement in the creation of whole-watershed management plans, runoff control programs may lack crucial public support and political momentum.

[40] House Committee on Public Works and Transportation, oversight hearing on the 208 program (1980), 16, 18.

[41] House Committee on Public Works and Transportation, oversight hearing on the 208 program (1980), 27–28.

[42] According to a longtime water policy activist with the League of Women Voters of the United States, there were "tens of thousands of meetings on 208 plans nationwide over a three-year period in the mid-1970s, and LWV members headed many of the 208 committees...virtually every local League was into the 208

process." Merilyn Reeves, former Board member, League of Women Voters of the United States, personal communication (May 15, 1992).

43 Paul Thompson, *Poison Runoff: A Guide to State and Local Control of Nonpoint Source Water Pollution* (NRDC, 1989), 21–22.

44 To quote from Thompson, Poison Runoff, chap. 2, p. 30, n.21: "[T]he degree to which poison runoff can be controlled dictates whether or not designated uses of individual waters are considered attainable: 'At a minimum, uses are deemed attainable if they can be achieved by the imposition of effluent limits...and cost-effective and reasonable best management practices for nonpoint source control.'" 40 CFR 131.10(d), 131.10(h)(2); 33 USC 1315(b)(1).

In effect, a state cannot legally decide that the minimum fishable/swimmable goal of the Clean Water Act is not attainable in a particular surface water unless the state has developed a poison runoff control program that controls nonpoint sources to the maximum extent practicable, and still is unable to achieve fishable/swimmable water quality. Similarly, under EPA's antidegradation regulation, even where water quality is better than necessary to protect designated instream uses, allowing further degradation is prohibited unless, among other requirements, the state assures the achievement of "all cost effective and reasonable best management practices for nonpoint source control." 40 CFR 131.12(a)(2); 33 USC 1313(e).

45 CWA § 305(b)(1)(E).

46 EPA, *Surface Water Monitoring: A Framework for Change* (1987), 4, iv, 27.

47 Claudia Copeland and Jeffrey A. Zinn, *Agricultural Nonpoint Pollution Policy: A Federal Perspective* (CRS, 1986), 8–11. EPA did try to pick up the slack in 208 funding via continued grants to states under CWA §§ 106 and 205(j). EPA, *National Water Quality Inventory, 1984*, EPA-440/4-85-029, 67.

48 Thompson, *Poison Runoff*, 22.

49 Thompson, *Poison Runoff*, 26.

50 Virginia Department of Conservation and Recreation, "Owl Run Livestock BMP Research Watershed" (fact sheet) (undated).

51 Kelly Allan, "One of the Last of the Best," in *The Nature Conservancy Magazine* (January/February 1991); Gary Overmier, U.S. Soil Conservation Service, personal communication, January 1, 1993.

52 George Hallberg et al., *A Progress Review of Iowa's Agricultural-Energy-Environmental Initiatives: Nitrogen Management in Iowa* (Iowa Department of Natural Resources, 1991), 5–6.

[53] EPA, *State Implementation of Nonpoint Source Programs,* draft report, Office of Policy, Planning and Evaluation (June 29, 1992), 7, 9, 11, 15, 18, 21, 24, 28, 31, 33, 34, 37.

[54] EPA, *State Implementation of Nonpoint Source Programs,* 39–41.

[55] EPA, *State Implementation of Nonpoint Source Programs,* 15.

[56] GAO, *Water Pollution: EPA Leadership Needed,* 28–29.

[57] We recognize that these criteria will need to be tailored to specific bioregions and basins; nonetheless, EPA guidance to the states, and a legislative mandate for adoption of such criteria, would be immensely beneficial.

[58] GAO, *Water Pollution: EPA Leadership Needed,* 14.

[59] CZARA § 6217(g)(5).

[60] EPA, *Guidance Specifying Management Measures for Sources of Nonpoint Pollution in Coastal Waters,* EPA Office of Water, 840-B-92-002 (1993).

[61] EPA, *Environmental Impacts of Stormwater Discharges: A National Profile,* EPA-841/R-92-001 (1992), 7.

[62] EPA, "Characteristics of Urban Runoff," in *Results of the Nationwide Urban Runoff Program* (1983), vol. 1, Final Report, chap. 6.

[63] *See,* for example, NRDC's summary of results for the four Chesapeake Bay cities, in R. Cohn-Lee and D. Cameron, "Urban Stormwater Runoff Contamination of the Chesapeake Bay: Sources and Mitigation," *The Environmental Professional* 14 (1992), 10–27.

[64] EPA, *Nationwide Urban Runoff Program,* 5-8 and 5-9.

[65] Metropolitan Washington Council of Governments, *Watershed Restoration Sourcebook,* Anacostia Restoration Team, Dept. of Environmental Programs, collected papers presented at the conference "Restoring Our Home River: Water Quality and Habitat in the Anacostia" (College Park, Md.: Nov. 6–7, 1991), 15–17.

[66] Bruce Lane and John Tanacredi, "Coastal Fisheries Project at Jamaica Bay Completed," *Park Science* (Winter 1987).

[67] Gateway National Recreation Area, *Shore Based Recreational Fishing Survey* (1985–86). The summer and winter flounders that are among the most popular eating fish from the Bay have average PCB concentrations around 0.1 to 0.2 parts/mil, roughly ten times less than EPA's recommended tolerance level of 2.0 parts/mil.

[68] 40 CFR pt. 122 and Apps. F–I.

[69] EPA, *Environmental Impacts of Stormwater Discharges,* 11.

[70] EPA, *Natural Resources for the 21st Century: An Evaluation of the Effects of Land Use on Environmental Quality, Office of Policy, Planning and Evaluation* (1989), 57.

[71] This has been shown to be quite feasible as a design principle for landscape architects and less costly by a factor of four than conventional pave-as-usual, treat-later, end-of-pipe approaches. Robert D. Sykes, "Site Planning," *Protecting Water Quality in Urban Areas, Best Management Practices for Minnesota*, Minnesota Water Pollution Control Agency (1989), chap. 3.1. The author, Sykes, ASLA, is Assoc. Prof. of Landscape Architecture, Univ. of Minnesota.

> The modern classic example of a comprehensive approach to development incorporating all of these [water-sensitive site design] goals is Woodlands New Community located north of Houston, Texas, planned and designed by Wallace, McHarg, Roberts and Todd, Landscape Architects and Planners, Philadelphia, Pennsylvania.... In the original planning, engineers compared the cost of the natural drainage system to that for a conventional approach and found that the natural drainage option saved over $14 million.

Id., 61, 3.1–7.

[72] NRDC, Water and Coastal Program, draft concept paper for "The Urban Watershed Restoration Act of 1992" (1992), 6 pp.

CHAPTER 7

[1] NRC, *Restoration of Aquatic Ecosystems: Science, Technology, and Public Policy* (Washington, D.C.: National Academy Press, 1992), 172.

[2] "Antidegradation Policy," 40 CFR § 131.12.

[3] NWF, "Waters At Risk: Keeping Clean Waters Clean" (1992), 4.

[4] NWF, "Waters at Risk," 14–15.

[5] NWF, "Waters at Risk," 13.

[6] NWF, "Waters at Risk," 25–27.

[7] 40 CFR § 131.12(a)(2).

[8] CWA § 401(a)(1).

[9] CWA § 401(a)(1). Conversely, "No license or permit shall be granted if certification has been denied by the State, interstate agency, or the Administrator, as the case may be." Id.

[10] CWA § 401(d).

[11] Roosevelt Campobello International Park Commission v. EPA, 684 F.2d 1041 (1st Cir. 1982).

NOTES

[12] U.S. v. Marathon Development Corp., 867 F.2d 96 (1st Cir. 1989).

[13] PUD No. 1 of Jefferson County and the City of Tacoma v. Depts. of Ecology, Fisheries and Wildlife, No. 58272-6, WA S.Ct. (April 1, 1993).

[14] Georgia-Pacific Corp. v. Vt. Dept. of Environmental Conservation, No. S473-89 WnC (1992). Georgia-Pacific has asked the U.S. Supreme Court to review this ruling.

[15] Arnold Irrigation District v. Dept. of Environmental Quality, 717 P.2d 1274 (Or. Ct. App. 1986). The CRS recently conducted a complete review of judicial decisions interpreting § 401. Memorandum from American Law Division, CRS, to House Committee on Interior and Insular Affairs, "Scope of State Authority to Condition or Deny Section 401 Certifications under the Clean Water Act: Review of State Case Law" (Dec. 21, 1992).

[16] K. Ransel and E. Myers, "State Water Quality and Wetland Protection: A Call to Awaken the Sleeping Giant," *Virginia Journal of Natural Resources Law* 7 (1988), 339.

[17] EDF v. Alexander, 501 F. Supp. 742 (N.D. Miss. 1980). § 401(a) provides that certification is waived if the state "fails or refuses to act on a request for certification, within a reasonable period of time (which shall not exceed one year) after receipt of such request."

[18] Ransel and Myers, "State Water Quality and Wetland Protection," 344 et seq.

[19] Ransel and Myers, "State Water Quality and Wetland Protection," 345.

[20] Ransel and Myers, "State Water Quality and Wetland Protection," 346.

[21] Ransel and Myers, "State Water Quality and Wetland Protection," 364–65, nn.101–6.

[22] Geoffrey H. Grubbs, Director, Assessment and Watershed Protection Division, EPA, personal communication (Oct. 20, 1992).

[23] Ransel and Myers, "State Water Quality and Wetland Protection," 343.

[24] Lake Erie Alliance for the Protection of the Coastal Corridor v. Army Corps of Engineers, 526 F. Supp. 1063, 1074 (W.D. Pa. 1981).

[25] NRDC v. EPA, 863 F.2d 1420, 1434-36 (9th Cir. 1988).

[26] Commonwealth of Pa., Dept. of Environmental Resources v. City of Harrisburg, 578 A.2d 563 (Pa.Cmwlth 1990).

[27] E.g., de Rham v. Diamond, 295 N.E. 2d 763 (N.Y. 1973); Power Authority of N.Y. v. Williams, 457 N.E. 2d 726 (N.Y. 1983). Other cases are discussed in the CRS analysis cited previously.

[28] CWA § 404(a) (emphasis added).

29 CWA § 404(b).

30 CWA § 404(c).

31 CWA § 404(f).

32 CWA § 404(e).

33 In fact, it took a lawsuit by NRDC to require the Corps to regulate wetlands at all under this provision. NRDC v. Callaway, 392 F.2d 685 (D.D.C. 1975). Activities in nonwetlands are also subject to regulation under § 10 of the Rivers and Harbors Act, which is beyond the scope of this study.

34 One problem with implementation of the program has been inconsistency among Corps district offices in interpreting what activities are subject to regulation.

35 GAO, *Wetlands: The Corps of Engineers' Administration of the Section 404 Program*, GAO/RCED-88-110 (1988), 19–20; OTA, U.S. Congress, *Wetlands: Their Use and Regulation*, OTA-0-206 (1984).

36 Dahl et al., *Status and Trends of Wetlands*, 2, 12–14.

37 GAO, *Wetlands Overview: Federal and State Policies, Legislation and Programs*, GAO/RCED-92-79FS (1991), 21.

38 OTA, *Wetlands: Their Use and Regulation*, 3–4.

39 EDF and WWF, *How Wet is a Wetland?*

40 NRDC v. Callaway, 392 F. Supp. 685 (D.D.C. 1975); U.S. v. Holland, 373 F. Supp. 665 (M.D. Fla. 1974).

41 U.S. v. Riverside Bayview Homes, 474 U.S. 121 (1985). The question of whether Congress may similarly regulate wetlands that are not associated with surface waters, however, remains as a cloud over the 404 program. Recently, a three-judge panel in the Seventh Circuit answered this question in the negative, but the opinion was vacated and will be reheard by the full Seventh Circuit Court of Appeals. Hoffman Homes v. EPA, No. 90-3810 (April 20, 1992), vacated, rehearing pending.

42 As one ecologist noted, "Wetlands are a half-way world between terrestrial and aquatic ecosystems and exhibit some of the characteristics of each." R. L. Smith (1980), quoted in Mitsch and Gosselink (1986), at 16.

43 33 CFR 328.3(b); 40 CFR 230.3(t).

44 Cowardin et al., *Classification of Wetlands and Deepwater Habitats in the U.S.*, FWS, U.S. Dept. of Interior, (1979).

45 William J. Mitsch and James G. Gosselink, *Wetlands* (N.Y.: Van Nostrand Reinhold, 1986), 18.

[46] Cowardin et al., *Classification of Wetlands and Deepwater Habitats,* 3.

[47] EDF and WWF, *How Wet Is a Wetland?* 11.

[48] William S. Sipple, "A Time to Move On," *National Wetlands Newsletter* 14, no. 2 (March/April 1992), 4–5.

[49] W. L. Want, *Law of Wetlands Regulation,* Environmental Law Series (release no. 3, May 1992), 4–29.

[50] For a brief description of some of these bills, *see* Jeffrey Zinn, CRS, Selected Wetlands Proposals Introduced in the 102d Congress (June 7, 1991).

[51] EDF and WWF, *How Wet is a Wetland?* x.

[52] "Evaluation of Proposed Revisions to the 1989 'Federal Manual for Identifying and Delineating Jurisdictional Wetlands,'" report of the Ecological Society of America's Ad Hoc Committee on Wetlands Delineation (Nov. 11, 1991).

[53] Sipple, "A Time to Move On," 5.

[54] GAO, *Wetlands Overview,* 32.

[55] 33 CFR, pt. 330, app. A.

[56] Want, *Law of Wetlands Regulation,* 5–8.

[57] Isolated wetlands are those that are not adjacent to another surface water, such as a river or lake. 33 C.F.R. 330.2(e).

[58] GAO, *Wetlands: The Corps of Engineers' Administration of the Section 404 Program,* GAO/RCED-88-110 (1988), 20–21.

[59] D. N. Gladwin and J. E. Roelle, "Nationwide Permits–Case Study Highlights Concerns," *National Wetlands Newsletter* 14, no. 2 (March/April 1992), 8–9.

[60] Gladwin and Roelle, "Nationwide Permits–Case Study Highlights Concerns," 9.

[61] 40 CFR 230.40–45.

[62] 40 CFR 230.10(a).

[63] 40 CFR 230.10(a)(3).

[64] GAO, *Wetlands: Corps of Engineers' Administration of Section 404 Program,* 26–28.

[65] GAO, *Wetlands: Corps of Engineers' Administration of Section 404 Program,* 27, quoting May 26, 1987, Memorandum from EPA Region VI to EPA Office of Wetlands Protection.

[66] 53 Fed. Reg. 3136 (Feb. 3, 1988), incorporated at 33 CFR 325, App. B(9)(c)(4).

[67] 40 CFR § 230.10(d). Specific ways to minimize impacts are included in 40 CFR 230, subpt. H.

[68] Want, *Law of Wetlands Regulation,* 6–27.

⁶⁹ Memorandum of agreement between the EPA and the Dept. of the Army concerning the determination of mitigation under the CWA § 404(b)(1) *Guidelines* (1990), pt. II.C.

⁷⁰ Conservation Foundation, *Protecting America's Wetlands: An Action Agenda*, The Final Report of the National Wetlands Policy Forum (1988), 3.

⁷¹ Conservation Foundation, *Protecting America's Wetlands*, 3.

⁷² EDF and WWF, *How Wet Is a Wetland?* xix.

⁷³ NRC, *Restoration of Aquatic Ecosystems*, 289–90.

⁷⁴ NRC, *Restoration of Aquatic Ecosystems*, 282–83.

⁷⁵ NRC, *Restoration of Aquatic Ecosystems*, 11–12.

⁷⁶ NRC, *Restoration of Aquatic Ecosystems*, 284–85.

⁷⁷ 40 CFR § 230.11(g).

⁷⁸ NRC, *Restoration of Aquatic Ecosystems*, 279.

⁷⁹ NRC, *Restoration of Aquatic Ecosystems*, 279–80.

⁸⁰ GAO, *Wetlands: Corps of Engineers' Administration of Section 404 Program*, 28–32.

⁸¹ GAO, *Wetlands: Corps of Engineers' Administration of Section 404 Program*, 37–47.

⁸² GAO, *Wetlands: Corps of Engineers' Administration of Section 404 Program*, 48.

⁸³ GAO, *Wetlands: Corps of Engineers' Administration of Section 404 Program*, 48.

⁸⁴ GAO, *Wetlands: Corps of Engineers' Administration of Section 404 Program*, 55.

⁸⁵ GAO, *Wetlands: Corps of Engineers' Administration of Section 404 Program*, 61; Ehorn, "U.S. Environmental Protection Agency's Enforcement Activities," in *Wetlands and River Corridor Management*, proceedings of a conference (1989), 450–52.

⁸⁶ EPA, *Report to Congress: Water Quality of the Nation's Lakes*, EPA-440/5-89-003 (1989), 3–4.

⁸⁷ NRC, *Restoration of Aquatic Ecosystems*, 91.

⁸⁸ NRC, *Restoration of Aquatic Ecosystems*, 72.

⁸⁹ NRC, *Restoration of Aquatic Ecosystems*, 149.

⁹⁰ NRC, *Restoration of Aquatic Ecosystems*, 6.

⁹¹ NRC, *Restoration of Aquatic Ecosystems*, 147.

⁹² OTA, *Wastes in Marine Environments*, 3, 7.

⁹³ EPA, *The National Estuary Program After Four Years, A Report to Congress* (1992), 1–4. Three of these estuaries are on the West Coast, four in the Gulf of Mexico, and the rest on the East Coast. Id.

[94] EPA, *National Estuary Program After Four Years*, 9; EPA, *National Water Quality Inventory, 1990*, 48.

[95] EPA, *National Estuary Program After Four Years*, i, 25.

[96] These criteria are in 40 CFR, pt. 125, subpt. M.

[97] EPA, *Report to Congress on Implementation of Section 403(c) of the Federal Water Pollution Control Act*, EPA-503/6-90-001 (1990), 5.

[98] EPA, *Report on Implementation of Section 403(c)*, 9.

[99] EPA, *Report on Implementation of Section 403(c)*, Fig. 5.

[100] EPA, *Report on Implementation of Section 403(c)*, 25–27.

[101] EPA, *Report on Implementation of Section 403(c)*, 29.

[102] 40 CFR § 125.123(c).

[103] EPA, *Report on Implementation of Section 403(c)*, v.

[104] Horton and Eichbaum, *Turning the Tide;* Colborn et al., *Great Lakes, Great Legacy?*

[105] *See,* for example, NWF, Canadian Institute for Environmental Law and Policy, *A Prescription for Healthy Great Lakes* (1991); Sierra Club, *A Great Lakes Federal Agenda for the 1990s: Implementing the Great Lakes Water Quality Agreement,* Sierra Club Great Lakes Federal Policy Project (Madison, Wis.: (1990).

[106] Colborn et al., *Great Lakes, Great Legacy?* xxiii, 3–5.

[107] Colborn et al., *Great Lakes, Great Legacy?* 4.

[108] IJC, *Sixth Biennial Report*, 2.

[109] IJC, *Sixth Biennial Report*, 24–30. Substances identified in the IJC report include synthetic chemicals (PCBs, PDT, dieldrin, toxaphene, mirex, and hexachlorobenzene), production by-products (dioxin and furans, benzo(a)pyrene, and hexachlorobenzene), and two metals (lead and mercury).

[110] IJC, *Sixth Biennial Report*, 31–34.

[111] NWF and CIELAP, *Prescription for Healthy Great Lakes,* 60–62; Sierra Club, *Great Lakes Federal Agenda for the 1990s,* 1–8.

[112] Chesapeake Bay Commission, *Annual Report to the General Assemblies of Maryland, Pennsylvania and Virginia* (1991), 1.

[113] Chesapeake Bay Commission, *1991 Annual Report,* 5–10.

[114] Chesapeake Bay Commission, *1991 Annual Report,* 31–34.

[115] Chesapeake Bay Commission, *1991 Annual Report,* 29.

[116] Chesapeake Bay Commission, *1991 Annual Report,* 25–31.

[117] NRC, *Restoration of Aquatic Ecosystems,* 148.

[118] NRC, *Restoration of Aquatic Ecosystems,* 171.

[119] NRC, *Restoration of Aquatic Ecosystems,* 208.

[120] NRC, *Restoration of Aquatic Ecosystems,* 172.

[121] NRC, *Restoration of Aquatic Ecosystems,* 184–85.

CHAPTER 8

[1] A copy of the agenda and more information about the network can be obtained by writing to Robyn Roberts, Outreach Coordinator, Clean Water Network, 1350 New York Ave., NW, Ste. 300, Washington, D.C. 20005, or call (202) 624-9357.

[2] For a copy and more information on Water Quality 2000, write to Water Quality 2000, c/o Water Environment Federation, 601 Wythe St., Alexandria, Va. 22314-1994, or call (800) 666-0206.

Acronyms

AFS ....Alternative Farming Systems

BAT..................Best Available Technology

BCF..................Bioconcentration Factor

BCT ....Best Conventional Technology

BLM..................Bureau of Land Management

BOD ....Biological Oxygen Demand

BMP..................Best Management Practices

BPJ..................Best Professional Judgment

CBC ....Chesapeake Bay Commission

CBE ....Citizens for a Better Environment

CBF..................Chesapeake Bay Foundation

CCMP ....Comprehensive Conservation and Management Plan

CDC..................Centers for Disease Control and Prevention

CEQ..................President's Council on Environmental Quality

CIELAP ....Canadian Institute for Environmental Law and Policy

CRS..................Congressional Research Service

CSO ....Combined sewer overflows

CWA ....Clean Water Act

CZARA ....Coastal Zone Act Reauthorization Amendments of 1990

CZMA..............Coastal Zone Management Act

DEQ..................Department of Environmental Quality

DOI..................Department of the Interior

EDF..................Environmental Defense Fund

EPA..................Environmental Protection Agency

ACRONYMS

ESA .... Endangered Species Act

FEMA .... Federal Emergency Management Agency

FERDS .... Federal Reporting Data System

FDA .... Food and Drug Administration

FWPCA .... Federal Water Pollution Control Administration

FWS .... Fish and Wildlife Service

GAO .... General Accounting Office

GIS .... Geographic Information System

GPO .... Government Printing Office

GS .... U.S. Geological Service

IJC .... International Joint Commission on the Great Lakes

ISTEA .... Intermodal Surface Transportation and Efficiency Act

NAS .... National Audubon Society

NASQAN .... National Stream Quality Accounting Network

NCBP .... National Contaminant Biomonitoring Program

NEP .... National Estuary Program

NMFS .... National Marine Fisheries Service

NOAA .... National Oceanic and Atmospheric Administration

NPDES .... National Pollution Discharge Elimination System

NRC .... National Research Council

NRDC .... Natural Resources Defense Council

NURP .... Nationwide Urban Runoff Program

NWF .... National Wildlife Federation

OCS .... Outer Continental Shelf

ONRW .... Outstanding National Resource Waters

OCPSF .... Organic Chemicals, Plastics and Synthetic Fibers

OPPE .... Office of Policy, Planning and Evaluation

ORSANCO .... Ohio River Valley Sanitation Commission

Acronyms

OTA...............Office of Technology and Assessment

PIRG..............Public Interest Research Group

POTW..............Publicly-Owned Treatment Works

RCRA..............Resource Conservation and Recovery Act

SCS ....Soil Conservaton Service

SELC..............Southern Environmental Law Center

SRF ....State Revolving Fund

TMDL..............Total maximum daily load

TNC...............The Nature Conservancy

TRI ....Toxics Release Inventory

USFS..............U.S. Forest Service

WPCF..............Water Pollution Control Federation

WQA...............Water Quality Act of 1987

WRI...............World Resources Institute

WWF ....World Wildlife Fund

Index

Academy of Natural Sciences, 20
Agriculture:
  economic savings from changes in
    agricultural practices, 174–75
  poison runoff from, 176–81
Alabama, 33, 128
Alternative Conservation Systems (ACSs), 192
American Fisheries Society, 61, 63–64
American Planning Association, 251
Anacostia River, 194–95, 196
Aquatic species and ecosystems:
  aquatic toxicity, 73–76
  biological integrity of, 59
  conclusion, 85–86
  CWA objectives as to, 7, 59
  endangered, threatened, and jeopardized
    species, 59–67
  extinction of species, 59, 62, 63
  fish kills, 69, 70–73, 177
  lost or damaged habitats, 76–85
    coastal habitat, 83–85
    floodplains, 80–81
    inland fisheries habitat, 77–78
    riparian habitat, 80–82
    water projects altering stream flow,
      effect of, 82–83, 246–47
    wetlands, 78–80, 85
  protection for, see Protection for
    aquatic resources and ecosystems
  trends in aquatic biodiversity, 61–67
  trends in populations of, 67–69, 7
  see also Fish and shellfish
Army Corps of Engineers, U.S., 8, 81, 94,
    98, 204, 205, 247–48
  permits for discharge of dredge or fill,
    see Permits, for discharge of dredge
    or fill
  wetlands delineation, 207, 208, 210
Ashland Oil, 168–69

Baker, Howard, 2
Baltimore Harbor, 56
Beaches:
  monitoring and reporting to protect
    public health, 132–35
  swimming hazards and beach closures,

see Swimming hazards and beach
  closures
Benke, Dr. Arthur C., 82–83
Best Available Technology (BAT)
  standards for toxic pollutants, 139,
    141–42, 143
  permits based on, 139–41
Best Conventional Technology (BCT)
  standards for conventional
    pollutants, permits based on, 139
Best Professional Judgment, permits based
  on, 139, 141, 164
Big Darby Creek Watershed, Ohio, 187
Big Spring Basin, Iowa, 187
Bioaccumulation, 43–44, 125
Bioconcentration, 43, 124, 125
Biological water quality criteria, 28–29, 120,
    125–26, 190, 229, 246
Biomagnification, 43–44, 125
Brandywine Creek, 5
Bureau of Land Management (BLM), 180,
    181, 242
Bureau of Sport Fisheries, U.S., 5
Bush, George, 213
Bush administration, 210
Buzzards Bay, 219

California:
  beach closures and advisories in, 32, 34
  coastal wetlands, 85
  sportfishing, 100
  wetlands, 211
California Safe Drinking Water and Toxic
  Enforcement Act, 122
Canadian Institute for Environmental Law
  and Policy, 74
Carson, Rachel, 6, 58
Centers for Disease Control and Prevention,
    38, 51–52
CEQ, see President's Council for
  Environmental Quality
Charles River, 196
Charles River watershed, wetlands in the, 94
Chesapeake Bay, 6, 9, 20, 21, 84, 167, 186, 194

311

INDEX